Praxis Study Guide for the *Mathematics* Tests

► ► ► ► ► ► ► ► ► ► ►

A PUBLICATION OF EDUCATIONAL TESTING SERVICE

Table of Contents

Praxis Study Guide for the *Mathematics* Tests

▶   ▶   ▶   ▶   ▶   ▶   ▶   ▶   ▶   ▶   ▶   ▶

Appendix A

Appendix B

Chapter 1

Introduction to the *Mathematics* Tests and Suggestions for Using This Study Guide

▶   ▶   ▶   ▶   ▶   ▶   ▶   ▶   ▶   ▶   ▶   ▶

Introduction to the *Mathematics* Tests

The *Mathematics* tests are designed for prospective secondary-school mathematics teachers. The tests are designed to reflect current standards for knowledge, skills, and abilities in mathematics. Educational Testing Service (ETS) works in collaboration with teacher educators, higher education content specialists, and accomplished practicing teachers in the field of mathematics to keep the tests updated and representative of current standards.

This guide covers three different *Mathematics* tests. One of the tests is *multiple-choice;* that is, it presents questions with several possible answers choices, from which you must choose the best answer and indicate your response on an answer sheet. Other tests are *constructed-response* tests; that is, you are asked to answer a question or group of questions by writing out your response. It is not accurate to call constructed-response tests *essay* tests, since your response will not be graded on the basis of how it succeeds as an essay. Instead, your constructed response will be graded on the basis of how well it demonstrates an understanding of the principles of mathematics and their appropriate application.

The following tests are covered in this guide:

Mathematics: Content Knowledge (0061)				
Length of Test	**Number and Format of Questions**	**Content Categories (Grouped) and Approximate Number of Questions**	**Approximate Number of Questions**	**Approximate Percentage of Questions**
120 minutes	50 multiple-choice questions **Note:** Graphing calculator is required	I. Arithmetic and Basic Algebra (6-8) Geometry (4-6) Trigonometry (2-4) Analytic Geometry (2-4)	17 total	34%
		II. Functions and Their Graphs (5-7) Calculus (5-7)	12 total	24%
		III. Probability and Statistics (3-5) Discrete Mathematics (3-5) Linear Algebra (3-5) Computer Science (2-4) Mathematical Reasoning and Modeling (5-7)	21 total	42%

Mathematics: Proofs, Models, and Problems, Part 1 (0063)		
Length of Test	**Number and Format of Questions**	**Content Categories**

60 minutes	4 constructed-response questions
	  I. 1 proof
	  II. 1 model
	  III. 2 problems
	
	Note: Graphing calculator is required.
		I. Arithmetic and Basic Algebra
		II. Geometry
		III. Analytic Geometry
		IV. Functions and Their Graphs
		V. Probability and Statistics (Without Calculus)
		VI. Discrete Mathematics

Mathematics: Pedagogy (0065)		
Length of Test	**Number and Format of Questions**	**Content Categories**
60 minutes	3 constructed-response questions	Planning, implementing, and assessing instruction

How to Use This Book

This book gives you instruction, practice, and test-taking tips to help you prepare for taking the *Mathematics* tests. In chapter 2 you will find a discussion of the Praxis Series™—what it is and how the tests are developed.

If you plan to take the *Mathematics: Content Knowledge* test, you should turn to chapters 3, 4, 5, 6, and 9 to

- review the topics likely to be covered on the test
- review fifteen approaches to solving math problems
- get tips on succeeding at multiple-choice tests
- take a practice test
- see the answers to the questions in the practice test, along with explanations of those answers

If you plan to take one or more constructed-response tests, you should turn to chapter 3 to review the topics likely to be covered on the test, chapter 5 for information on how to succeed on this type of test, and chapter 4 to review approaches to solving math problems. Then chapters 7 and 10 (for the *Proofs, Models, and Problems, Part 1,* test) and chapters 8 and 11 (for the *Pedagogy* test) will help you prepare for the test and allow you to take a practice test and see sample responses and how they were scored.

So where should you start? Well, all users of this book will probably want to begin with the following two steps:

Become familiar with the test content. Note what the appropriate chapter of the book says about the topics covered in the test you plan to take.

Consider how well you know the content in each subject area. Perhaps you already know that you need to build up your skills in a particular area. If you're not sure, skim over the chapters that cover test content to see what topics they cover. If you encounter material that feels unfamiliar or difficult, fold down page corners or insert sticky notes to remind yourself to spend extra time in these sections.

Also, all users of this book will probably want to end with these two steps:

Familiarize yourself with test taking. Chapter 5 answers frequently-asked questions about multiple-choice tests, such as whether it is a good idea to guess on a test, and also explains how constructed-response tests are scored, with valuable tips on how to succeed on a test in this format. With either test format, you can simulate the experience of the test by taking a practice test within the specified time limits. Choose a time and place where you will not be interrupted or distracted. For a multiple-choice test, use the appropriate chapter to score your responses. The scoring key identifies which topic each question addresses, so you can see which areas are your strongest and weakest. Look over the explanations of the questions you missed and see whether you understand them and could answer similar questions correctly. Then plan any additional studying according to what you've learned about your understanding of the topics and your strong and weak areas. For a constructed-response test, you can see sample responses that scored well, scored poorly, or scored in-between. By examining these sample responses, you can focus on the aspects of your own practice response that were successful and unsuccessful. This knowledge will help you plan any additional studying you might need.

Register for the test and consider last-minute tips. Consult the Praxis Web site at www.ets.org/praxis to learn how to register for the test, and review the checklist in chapter 12 to make sure you are ready for the test.

What you do between these first steps and these last steps depends on whether you intend to use this book to prepare on your own or as part of a class or study group.

Using this book to prepare on your own:

If you are working by yourself to prepare for a *Mathematics* test, you may find it helpful to fill out the Study Plan Sheet in appendix A. This work sheet will help you to focus on what topics you need to study most, identify materials that will help you study, and set a schedule for doing the studying. The last item is particularly important if you know you tend to put off work.

Using this book as part of a study group:

People who have a lot of studying to do sometimes find it helpful to form a study group with others who are preparing toward the same goal. Study groups give members opportunities to ask questions and get detailed answers. In a group, some members usually have a better understanding of certain topics, while others in the group may be better at other topics. As members take turns explaining concepts to each other, everyone builds self-confidence. If the group encounters a question that none of the members can answer well, the members can go as a group to a teacher or other expert and get answers efficiently. Because study groups schedule regular meetings, group members study in a more disciplined fashion. They also gain emotional support. The group should be large enough so that various people can contribute various kinds of knowledge, but small enough so that it stays focused. Often, three to six people is a good size.

Here are some ways to use this book as part of a study group:

Plan the group's study program. Parts of the Study Plan Sheet in appendix A can help to structure your group's study program. By filling out the first five columns and sharing the work sheets, everyone will learn more about your group's mix of abilities and about the resources (such as textbooks) that members can share with the group. In the sixth column ("Dates planned for study of content"), you can create an overall schedule for your group's study program.

Plan individual group sessions. At the end of each session, the group should decide what specific topics will be covered at the next meeting and who will present each topic. Use the topic headings and subheadings in the chapter that covers the topics for the test you will take.

Prepare your presentation for the group. When it's your turn to be presenter, prepare something that's more than a lecture. Write five to ten original questions to pose to the group. Practicing writing actual questions can help you better understand the topics covered on the test as well as the types of questions you will encounter on the test. It will also give other members of the group extra practice at answering questions.

Take the practice test together. The idea of the practice test is to simulate an actual administration of the test, so scheduling a test session with the group will add to the realism and will also help boost everyone's confidence.

Learn from the results of the practice test. For each test, use the corresponding chapter with the correct answers to score each other's answer sheets. Then plan one or more study sessions based on the questions that group members got wrong. For example, each group member might be responsible for a question that he or she got wrong and could use it as a model to create an original question to pose to the group, together with an explanation of the correct answer modeled after the explanations in this study guide.

Whether you decide to study alone or with a group, remember that the best way to prepare is to have an organized plan. The plan should set goals based on specific topics and skills that you need to learn, and it should commit you to a realistic set of deadlines for meeting these goals. Then you need to discipline yourself to stick with your plan and accomplish your goals on schedule.

Chapter 2
Background Information on
The Praxis Series™ Assessments

▶ ▶ ▶ ▶ ▶ ▶ ▶ ▶ ▶ ▶ ▶ ▶

What Are The Praxis Series™ Subject Assessments?

The Praxis Series™ Subject Assessments are designed by Educational Testing Service (ETS) to assess your knowledge of the subject area you plan to teach, and they are a part of the licensing procedure in many states. This study guide covers an assessment that tests your knowledge of the actual content you hope to be licensed to teach. Your state has adopted The Praxis Series tests because it wants to be certain that you have achieved a specified level of mastery of your subject area before it grants you a license to teach in a classroom.

The Praxis Series tests are part of a national testing program, meaning that the test covered in this study guide is used in more than one state. The advantage of taking Praxis tests is that if you want to move to another state that uses The Praxis Series tests, you can transfer your scores to that state. Passing scores are set by states, however, so if you are planning to apply for licensure in another state, you may find that passing scores are different. You can find passing scores for all states that use The Praxis Series tests in the *Understanding Your Praxis Scores* pamphlet, available either in your college's School of Education or by calling (609) 771-7395.

What Is Licensure?

Licensure in any area—medicine, law, architecture, accounting, cosmetology—is an assurance to the public that the person holding the license has demonstrated a certain level of competence. The phrase used in licensure is that the person holding the license *will do no harm*. In the case of teacher licensing, a license tells the public that the person holding the license can be trusted to educate children competently and professionally.

Because a license makes such a serious claim about its holder, licensure tests are usually quite demanding. In some fields licensure tests have more than one part and last for more than one day. Candidates for licensure in all fields plan intensive study as part of their professional preparation: some join study groups, others study alone. But preparing to take a licensure test is, in all cases, a professional activity. Because it assesses your entire body of knowledge or skill for the field you want to enter, preparing for a licensure exam takes planning, discipline, and sustained effort. Studying thoroughly is highly recommended.

Why Does My State Require The Praxis Series Assessments?

Your state chose The Praxis Series Assessments because the tests assess the breadth and depth of content— called the "domain" of the test—that your state wants its teachers to possess before they begin to teach. The level of content knowledge, reflected in the passing score, is based on recommendations of panels of teachers and teacher educators in each subject area in each state. The state licensing agency and, in some states, the state legislature ratify the passing scores that have been recommended by panels of teachers.

You can find out the passing score required for The Praxis Series Assessments in your state by looking in the pamphlet *Understanding Your Praxis Scores*, which is free from ETS (see above). If you look through this pamphlet, you will see that not all states use the same test modules, and even when they do, the passing scores can differ from state to state.

What Kinds of Tests Are The Praxis Series Subject Assessments?

Two kinds of tests comprise The Praxis Series Subject Assessments: multiple choice (for which you select your answer from a list of choices) and constructed response (for which you write a response of your own). Multiple-choice tests can survey a wider domain because they can ask more questions in a limited period of time. Constructed-response tests have far fewer questions, but the questions require you to demonstrate the depth of your knowledge in the area covered.

What Do the Tests Measure?

The Praxis Series Subject Assessments are tests of content knowledge. They measure your understanding of the subject area you want to teach. The multiple-choice tests measure a broad range of knowledge across your content area. The constructed-response tests measure your ability to explain in depth a few essential topics in your subject area. The content-specific pedagogy tests, most of which are constructed-response, measure your understanding of how to teach certain fundamental concepts in your field. The tests do not measure your actual teaching ability, however. They measure your knowledge of your subject and of how to teach it. The teachers in your field who help us design and write these tests, and the states that require these tests, do so in the belief that knowledge of subject area is the first requirement for licensing. Your teaching ability is a skill that is measured in other ways: observation, videotaped teaching, or portfolios are typically used by states to measure teaching ability. Teaching combines many complex skills, only some of which can be measured by a single test. The Praxis Series Subject Assessments are designed to measure how thoroughly you understand the material in the subject areas in which you want to be licensed to teach.

How Were These Tests Developed?

ETS began the development of The Praxis Series Subject Assessments with a survey. For each subject, teachers around the country in various teaching situations were asked to judge which knowledge and skills a beginning teacher in that subject needs to possess. Professors in schools of education who prepare teachers were asked the same questions. These responses were ranked in order of importance and sent out to hundreds of teachers for review. All of the responses to these surveys (called "job analysis surveys") were analyzed to summarize the judgments of these professionals. From their consensus, we developed the specifications for the multiple-choice and constructed-response tests. Each subject area had a committee of practicing teachers and teacher educators who wrote these specifications (guidelines). The specifications were reviewed and eventually approved by teachers. From the test specifications, groups of teachers and professional test developers created test questions.

When your state adopted The Praxis Series Subject Assessments, local panels of practicing teachers and teacher educators in each subject area met to examine the tests question by question and evaluate each question for its relevance to beginning teachers in your state. This is called a "validity study." A test is considered "valid" for a job if it measures what people must know and be able to do on that job. For the test to be adopted in your state, teachers in your state must judge that it is valid.

These teachers and teacher educators also performed a "standard-setting study"; that is, they went through the tests question by question and decided, through a rigorous process, how many questions a beginning teacher should be able to answer correctly. From this study emerged a recommended passing score. The final passing score was approved by your state's Department of Education.

In other words, throughout the development process, practitioners in the teaching field—teachers and teacher educators—have determined what the tests would contain. The practitioners in your state determined which tests would be used for licensure in your subject area and helped decide what score would be needed to achieve licensure. This is how professional licensure works in most fields: those who are already licensed oversee the licensing of new practitioners. When you pass The Praxis Series Subject Assessments, you and the practitioners in your state can be assured that you have the knowledge required to begin practicing your profession.

Chapter 3
Study Topics

▶ ▶ ▶ ▶ ▶ ▶ ▶ ▶ ▶ ▶ ▶ ▶

This chapter and the next chapter ("Fifteen Approaches to Solving Math Problems") are intended to help you organize your preparation for the Praxis *Mathematics* tests and to give you a clear indication about the depth and breadth of the knowledge required for success on the tests. In this chapter, the *Content Knowledge* and *Proofs, Models, and Problems, Part 1* tests are covered first, followed by the *Pedagogy* test.

You are not expected to be an expert on all aspects of the knowledge and skills statements that follow. You should understand the major concepts and procedures associated with each statement. Virtually all accredited undergraduate mathematics programs address the majority of these topics.

When you find a skill or topic that is unfamiliar or fuzzy to you, you'll need to find out more. Consult materials and resources, including lecture and seminar notes, from all your mathematics course work. You should be able to match up specific topics with what you have covered in your courses. You may also, at times, want to refer to supplementary books or Web sites that cover the material. In addition, you should seek assistance from a professor or mentor teacher if you are stuck.

Try not to be overwhelmed by the volume and scope of knowledge and skills in this guide. An overview such as this does not offer you a great deal of context. Although a specific term may not seem familiar as you see it here, you might find you can understand it when applied to a real-life situation. Many of the items on the actual Praxis test will provide you with a context in which to apply these topics or terms, as you will see when you look at the practice tests.

The *Content Knowledge* Test and the *Proofs, Models, and Problems, Part 1* Test

Questions on the *Mathematics: Content Knowledge* and *Proofs, Models, and Problems, Part 1* tests can all be solved by using skills and abilities from twelve different areas. A summary of the skills and abilities required in each of these areas is listed below. These skills and abilities are required for both tests unless otherwise indicated.

A. Arithmetic and Basic Algebra

- Understand and work with rational, irrational, real, and/or complex numbers. Use numbers in a way that is most appropriate in the context of a problem (e.g., appropriately rounded numbers, numbers written in scientific notation, using $5(100 - 1)$ for $5(99)$, etc.).
- Demonstrate understanding of the properties of counting numbers (e.g., prime or composite, even or odd, factors, multiples).
- Apply the order of operations to problems involving addition, subtraction, multiplication, division, roots, and powers, with and without grouping symbols.
- Identify the properties (e.g., closure, commutativity, associativity, distributivity) of the basic operations (i.e., addition, subtraction, multiplication, division) on the standard number systems.
- Given newly defined operations on a number system, determine whether the closure, commutative, associative, or distributive properties hold.
- Identify the additive and multiplicative inverses of a number.
- Interpret and apply the concepts of ratio, proportion, and percent in appropriate situations.

- Solve problems that involve measurement in the metric or the traditional system.
- Solve problems involving average, including arithmetic mean and weighted average.
- Work with algebraic expressions and formulas.
  - Example: If $x = 5$ and $y = 6$, what is the value of $x^2 + 5y$? If $A = \dfrac{Bh}{2}$, then express h in terms of A and B.
- Add, subtract, multiply, and divide polynomials, as well as algebraic fractions.
- Translate verbal expressions and relationships into algebraic expressions or equations.
- Solve and graph linear equations and inequalities in one or two variables; solve and graph systems of linear equations and graph inequalities in two variables; solve and graph nonlinear algebraic equations and graph inequalities.
- Determine any term of a binomial expansion using Pascal's triangle or some other method.
  - Example: What is the fourth term of the binomial expansion of $(a + b)^5$?
- Solve equations and inequalities involving absolute values.
  - Examples: $|x + 2| = 6$, $\ |x + 2| + 3x = 6$
- Interpret and present geometric interpretations of algebraic principles (e.g., the triangle inequality and the distributive principles).

B. Geometry

- Solve problems involving the properties of parallel and perpendicular lines.
- Solve problems using special triangles, such as isosceles and equilateral.
- Solve problems using the relationships of the parts of triangles, such as sides, angles, medians, midpoints, and altitudes.

- Apply the Pythagorean theorem to solve problems.
- Solve problems using the properties of special quadrilaterals, such as the square, rectangle, parallelogram, rhombus, and trapezoid, and describe relationships among these sets of special quadrilaterals.
- Solve problems using the properties (e.g., angles, sum of angles, number of diagonals, and vertices) of polygons with more than four sides.
- Solve problems using the properties of circles, including those involving inscribed angles, central angles, chords, radii, tangents, secants, arcs, and sectors.
- Compute the perimeter and area of triangles, quadrilaterals, and circles, and of regions that are combinations of these figures.
- Use relationships (e.g., congruency, similarity) among two-dimensional geometric figures and among three-dimensional figures to solve problems.
- Compute the surface area and volume of right prisms, pyramids, cones, cylinders, and spheres, and of solids that are combinations of these figures.
- Solve problems involving reflections, rotations, and translations of points, lines, or polygons in the plane.
- Execute geometric constructions using a straightedge and compass
  - Examples: Bisect an angle, construct a perpendicular, prove that a construction yields the desired result.

C. Trigonometry

- Identify the relationship between radian measures and degree measures of angles.
- Define and use the six basic trigonometric relationships in the context of a right triangle and, using radian measure, in the context of the unit circle.

- Solve problems involving right triangles and problems involving trigonometric functions evaluated at such numbers as $\pi, \frac{\pi}{6}, \frac{2\pi}{3}$, $\frac{9\pi}{4}$, and $-\frac{9\pi}{3}$.

- Apply the law of sines and the law of cosines in the solution of problems. Recognize the graphs of the six basic trigonometric functions and identify their period, amplitude, phase displacement or shift, and asymptotes.

- Apply the formulas for the trigonometric functions of $\frac{x}{2}, 2x, x+y,$ and $x-y$ in terms of the trigonometric functions of x and y.

- Prove identities using the basic trigonometric identities.

- Solve trigonometric equations and inequalities.

- Given a point in the rectangular coordinate system, identify the corresponding point in the polar coordinate system.

- Find the trigonometric form of complex numbers and apply De Moivre's theorem.

D. Functions and Their Graphs

- Understand function notation for the functions of one variable, and be able to work with the algebraic definition of a function (i.e., for every x there is at most one y), and be able to identify whether a graph in the plane is a graph of a function.

- Use multiple representations of a function, such as an equation, a graph, a table, or a verbal statement.

- Use the definition of a function as a mapping and be able to work with functions given in this way.

  ○ Example: $f:(x,\ y) \rightarrow \left(x^2 + y^2,\ x^2 - y^2\right)$

- Find the domain and/or range of a function.

- Use the properties of algebraic, trigonometric, logarithmic, and exponential functions to solve problems (e.g., finding composite functions and inverse functions).

- Find the inverse of a one-to-one function in simple cases and know why one-to-one functions have inverses.

- Determine the graphical properties and sketch a graph of a linear, step, absolute-value, or quadratic function (e.g., slope, intercepts, intervals of increase or decrease, axis of symmetry).

E. Probability and Statistics

- Organize data into a presentation that is appropriate for solving a problem.
  ○ Example: construct a histogram and use it to calculate probabilities

- Solve probability problems involving finite sample spaces by actually counting outcomes appropriately.

- Solve probability problems by using counting techniques.
  ○ Example: If 3 cards are drawn from a standard deck of 52 cards, what is the probability that all 3 will be aces?

- Solve probability problems involving independent trials.
  ○ Example: If a coin is tossed 5 times, what is the probability that heads will occur at least 3 times?

- Solve problems by using the binomial distribution and be able to determine when the use of the binomial distribution is appropriate.

- Solve problems involving joint probability.

- Find and interpret common measures of central tendency (population mean, sample mean, median, mode) and know which is the most meaningful to use in a given situation.

- Find and interpret common measures of dispersion (range, population standard deviation, sample standard deviation, population variance, sample variance).
- Model an applied problem by using the mathematical expectation of an appropriate discrete random variable (e.g., fair coins, expected winnings, expected profit).
- Solve problems using the normal distribution.
- Solve basic intuitive problems using the concepts of uniform and chi-square distributions (no technical vocabulary).
- Recognize a valid test to determine whether to accept or reject a given null hypothesis, H_0.

F. Analytic Geometry

- Determine equations of lines and planes, given appropriate information.
- Make calculations in 2-space or 3-space (e.g., distance between two points, coordinates of the midpoint of a line segment, distance between a point and a plane).
- Given a geometric definition of a conic section, derive the equation for the conic section.
  - Example: Given that a parabola is the set of points that are equidistant from a given point and a given line, derive its equation.
- Determine which conic section is represented by a given equation if the axis of symmetry is parallel to one of the coordinate axes and if no restrictions are placed on its location in the plane.

G. Calculus

- Discuss informally what it means for a function to have a limit at a point.
- Calculate limits of functions or determine that the limit does not exist.

- Solve problems using the properties of limits.
  - Example:
    $$\lim_{x \to c} f(x) + \lim_{x \to c} g(x) = \lim_{x \to c} (f(x) + g(x))$$
- Use limits to show that a particular function is continuous.
- Use L'Hôpital's rule, where applicable, to calculate limits of functions.
- Relate the derivative of a function to a limit or to the slope of a curve.
- Explain conditions under which a continuous function does not have a derivative.
- Differentiate algebraic expressions, trigonometric functions, and exponential and logarithmic functions using the sum, product, quotient, and chain rules.
- Use implicit differentiation.
- Make numerical approximations of derivatives and integrals.
- Use differential calculus to analyze the behavior of a function (e.g., find relative maxima and minima, concavity).
- Use differential calculus to solve problems involving related rates and rates of change.
- Approximate the roots of a function (e.g., using Newton's method with derivatives).
- Use differential calculus to solve applied minima-maxima problems.
- Solve problems using the Mean Value Theorem of differential calculus.
- Explain the significance of, and solve problems using, the fundamental theorem of calculus.
- Demonstrate an intuitive understanding of the process of integration as finding areas of regions in the plane through a limiting process.
- Integrate functions using standard integration techniques.
- Evaluate improper integrals.
- Use integral calculus to calculate the area of regions in the plane and the volume of solids formed by rotating plane figures about a line.

- Determine the limits of sequences and simple infinite series.
- Use standard tests to show convergence (either conditional or absolute) or divergence of series (e.g., comparison, ratio).

H. Discrete Mathematics

- Use the basic terminology and, given the definitions, use the symbols of logic.
- Use truth tables to verify statements.
- Use laws of Algebra of Propositions to evaluate equivalence of complex logical expressions (e.g., De Morgan's laws).
- Solve problems involving the union and intersection of sets, subsets, and disjoint sets.
- Solve basic problems involving permutations and combinations.
- Use the Euclidean algorithm to find the greatest common divisor of two numbers.
- Work with numbers expressed in bases other than base 10.
- Find values of functions defined recursively and "translate" between recursive and closed-form expressions for a function.
- Determine whether a binary relation on a set is reflexive, symmetric, antisymmetric, transitive, or an equivalence relation.
- Solve simple linear programming problems.

I. Abstract Algebra

- Determine whether a particular set together with a given operation is a group.
- Determine whether a particular set together with two operations is a field.

J. Linear Algebra

- Add, subtract, and scalar multiply vectors, using geometric interpretations of these operations, and use in real-world applications.
- Scalar multiply, add, subtract, and multiply matrices.

- Demonstrate an understanding and use the basic properties of inverses of matrices.
- Determine and apply the matrix representation of a linear transformation.
- Use matrix techniques to solve systems of linear equations.

K. Computer Science

- Demonstrate an understanding of the roles of the hardware and software components of a computer system (e.g., output devices, CPU, disks, operating systems, secondary storage devices).
- Know basic computer terminology (e.g., files, I/O, records).
- Be able to use "user-friendly" software (e.g., classroom instruction packages, graphics software, spreadsheets).
- Develop computer algorithms to solve mathematical problems.
- Trace and debug existing computer algorithms.

L. Mathematical Reasoning and Modeling

- Demonstrate an understanding of a physical situation or a verbal description of a situation, develop a mathematical model of it, and determine whether one mathematical model will describe two apparently different situations.
- Determine appropriate mathematical strategies to solve a problem. These strategies might include conjectures, counterexamples, inductive reasoning, deductive reasoning (mathematical induction, proof by contradiction, direct proof, and other types of proof), and deciding which tools are appropriate (e.g., discussion with others, mental math, pencil and paper, calculator, computer, trees and graphs, fingers).

- Recognize the reasonableness of results, given the context of a problem.
- Using estimation, test the reasonableness of results.
- Estimate the actual and relative error in the numerical answer to a problem by analyzing the effects of round-off and truncation errors introduced in the course of solving a problem.
- Having solved a problem, reconsider the strategies used. Are there other appropriate strategies? Which strategies are the most efficient? Can these strategies be used to solve other problems? Can these strategies be used to prove a more general result?
- Communicate results in an appropriate form (e.g., correct English sentences, tables, charts, graphs).
- Demonstrate an understanding of the different levels of mathematical impossibility, such as: I lack the mathematical skills to do it; no one has been able to do it as yet (e.g., prove Goldbach's conjecture); no one will ever be able to do it (e.g., trisect a general angle with straightedge and compass).
- Use the axiomatic method in modeling and problem solving.

The *Pedagogy* Test

The *Mathematics: Pedagogy* test does not require as high a level of mathematical knowledge and skill as the *Content Knowledge* or *Proofs, Models, and Problems, Part 1* test. In fact, the level of mathematics tested does not go beyond first-year algebra. The focus of the *Pedagogy* test is on your ability to plan instruction, plan how to implement instruction, and plan appropriate methods of assessment and evaluation. Knowledge of *how* students learn mathematics is as important as knowledge *of* mathematics.

Each question on the *Mathematics: Pedagogy* test is based on one or more of the following knowledge or skills areas.

- Evaluate a given scope and sequence of mathematics topics.
- Integrate concepts to show relationships among topics.
- Develop a scope and sequence for a mathematics topic at a particular level and justify your choices.
- Given an example of a student's work that contains an error arising from a misconception, identify the misconception and suggest methods for correcting it.
- Identify the knowledge and skills that students need to master before being taught a particular topic.
- Develop questions that ask students to display their current level of understanding of a particular topic.
- Given a particular problem, identify several problem-solving strategies (e.g., guess and check, reduce to a simpler problem, draw a diagram, work backwards) that might help students solve the problem.
- Use appropriate forms of representation (e.g., analogies, drawings, examples, symbols, manipulatives) for mathematics subject matter for a particular group of students to help make mathematics understandable and interesting.
- Use a variety of teaching strategies (e.g., laboratory work, supervised practice, group work, lecture) in mathematics appropriate for a particular group of students and a particular topic.
- Relate mathematical concepts and ideas to real-world situations.
- Identify, evaluate, and use curricular materials and resources for mathematics (e.g., textbooks and other printed material, computer

software, base-10 blocks, geoboards, egg cartons) in ways appropriate for a particular group of students and a particular topic.

- Know procedures for controlling the social atmosphere of a classroom without restricting divergent mathematical thought.

- Know how society is affected by its general level of mathematical knowledge and know how societal influences differentially affect the mathematics education of gender, racial, ethnic, and socioeconomic groups.

- Know how to use information about different gender, ethnic, and socioeconomic groups to enhance these groups' learning of mathematics.

- Identify, evaluate, and use appropriate evaluation strategies (e.g., observations, interviews, oral discussions, written tests) to assess student progress in mathematics.

- Write specific evaluation items to test for a specific mathematical skill.

Chapter 4
Fifteen Approaches to Solving Math Problems

▶   ▶   ▶   ▶   ▶   ▶   ▶   ▶   ▶   ▶   ▶   ▶

Fifteen Approaches to Solving Math Problems

Solving a mathematics problem often requires more than understanding mathematical facts. It requires knowing *what* mathematical facts to use and *how* to use those facts to develop a solution to the problem.

Mathematics problems are solved by using a wide variety of strategies and techniques. Often there are several ways to solve a problem, all of which are correct. Experienced problem solvers have a repertoire of problem-solving strategies and techniques that they use frequently, as well as a sense of which are likely to work in solving a particular problem. This chapter focuses on a discussion of some of these strategies and techniques.

Unfortunately, when you must choose an appropriate strategy, there is no set of rules applicable to all problems. The ability to select a strategy grows as you solve more and more problems. However, no matter how experienced you become, there will be times when choosing an appropriate strategy will be difficult.

This chapter contains descriptions of fifteen strategies and techniques that are often useful in solving problems, as well as examples of problems for which they are appropriate. Read the descriptions of each strategy or technique and the associated examples, and try to think of other problems for which the strategy or technique might be useful. When you solve problems, it is important that your thinking remain flexible and that you keep an open mind. If your chosen strategy is not working, it may be useful to consider another approach. The strategies and techniques presented in this section are not presented in any particular order, nor are they a complete list.

1. Generating a counterexample

Mathematics often deals with assertions, which may be true or false. Various strategies may be used to prove that an assertion is true or that it is false. One way to prove that an assertion is false is to generate a counterexample. A counterexample is an example that satisfies the conditions of the assertion but does NOT satisfy the conclusion of the assertion. If you are successful in generating a counterexample, you have shown that the assertion is not true. If you are unable to generate a counterexample, it does NOT mean that you have shown that the assertion is true. In fact, the assertion may not be true, but you may just have been unsuccessful in generating a counterexample. To show that the assertion is true, you must show that the result is true in ALL cases in which the conditions are met.

Example 1: Suppose that you are given a polynomial of the form $P(x) = ax^2 + bx + c$, where a, b, and c are constants. If $P(1) = 0$ and $P(2) = 0$, does $P(x) = 0$ for all real numbers x?

In this problem, you are given a general polynomial of, at most, second degree and the values of that polynomial at two specific values of x. You are asked to determine whether the 0 function is the only function meeting those conditions. Clearly, the strategy of finding a counterexample is appropriate to try in solving the problem. The counterexample must be a linear or quadratic function that is 0 when $x = 1$ and when $x = 2$. Recall that the graph of a quadratic function is a parabola, so the counterexample would be a function whose graph is a parabola with zeros at $x = 1$ and $x = 2$. The function $P(x) = x^2 - 3x + 2 = (x-1)(x-2)$ meets these conditions, but $P(x) \neq 0$ for all real numbers x. (In fact $P(x)$ is equal to zero only when $x = 1$ or $x = 2$.)

Example 2: Suppose that you are asked whether every function $f(x)$ that is continuous at all real numbers x is also differentiable at all real numbers x.

In this problem, you are given that a function is continuous at all real numbers x and are asked to determine whether it must be true that it is differentiable at all real numbers x. Since this is a general statement, it is reasonable to try to solve the problem by generating a counterexample. In this case, a counterexample could be a function $f(x)$ that is continuous at all real numbers x but not differentiable at one real number. Once you have determined that this would provide a suitable counterexample, you may recall that the function $f(x) = |x|$ is continuous at all real numbers x but not differentiable at $x = 0$. This shows that if a function $f(x)$ is continuous at all real numbers x, it need not be differentiable at all real numbers.

2. Representing an algebraic problem geometrically

When you are given an algebraic problem, it is sometimes useful to represent the problem geometrically. Sometimes the geometric figure clarifies behavior that is obscure in the algebraic presentation.

Example 1: If $|x| + |x - 2| = 2$, where x is a real number, what is the value of x?

Often you can solve a problem involving the absolute value of an expression or equation with one variable by representing the problem geometrically. To use this strategy, recall the following two facts.

- The absolute value of a number r can be viewed as the distance, on the number line, of the point with coordinate r from the point with coordinate 0.
- The absolute value of a number $r - s$ is the distance, on the number line, of the point with coordinate r from the point with coordinate s.

Using these two facts, you can see that the problem is equivalent to the following geometric problem: find the number x such that [the distance of the point with coordinate x from the point with coordinate 0] plus [the distance of the point with coordinate x from the point with coordinate 2] is equal to 2.

Drawing a number line with this information on it gives

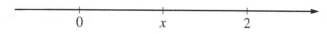

Inspection of the number line shows that $0 \le x \le 2$ is the solution.

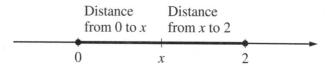

Example 2: If x and y are positive numbers and $x \leq y$, show that $\sqrt{xy} \leq \dfrac{x+y}{2}$.

A good approach for establishing some inequalities is to appeal to geometric interpretation. The given problem is not an exception to this rule. Notice that $\dfrac{x+y}{2}$ is the arithmetic mean of x and y and that $\sqrt{xy}$ is their geometric mean. Geometrically, in a right triangle, $\sqrt{xy}$ is the length of the altitude of the right triangle with hypotenuse $x+y$, and $\dfrac{x+y}{2}$ is the radius of the semicircle in which the right triangle can be inscribed. Therefore, you can consider the following figure.

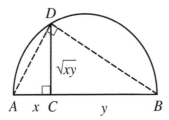

As you can see, the semicircle is constructed with diameter AB of length $x+y$, and C is a point chosen so that $AC = x$ and $CB = y$. A perpendicular erected from AB at C meets the circle at D. By construction, ADB, ACD, and DCB are right triangles. Now you can recall that the altitude to the hypotenuse of a right triangle forms two right triangles that are similar to each other. Therefore, you can conclude that triangles ACD and DCB are similar. Thus $\dfrac{x}{CD} = \dfrac{CD}{y}$ and therefore $CD = \sqrt{xy}$.

You can clearly see that $\sqrt{xy}$ is less than or equal to the radius of the semicircle, which equals $\dfrac{x+y}{2}$ (i.e., $\sqrt{xy} \leq \dfrac{x+y}{2}$).

3. Using proof by contradiction or indirect proof

Proof by contradiction (also called indirect proof) is a very powerful strategy. To use proof by contradiction, you begin by assuming that the conclusion you are supposed to reach is FALSE and proceed to show that this assumption leads to a result that CANNOT be true. The fact that you have been led to a result that cannot be true means that your assumption (that leads to the conclusion you were supposed to reach) is false, and the conclusion you were supposed to reach, therefore, must be TRUE.

Example 1: Prove that if a convex 9-sided polygon has an axis of symmetry, then the axis of symmetry passes through exactly one of the vertices of the polygon.

Often problems that can be solved by using proof by contradiction are problems involving "if … then" statements. In such cases, proof by contradiction is begun by assuming that the "then" statement is false.

In this case, you could begin either by assuming that the axis of symmetry did not pass through a vertex or that it passed through two vertices. In either of these cases, you can show that the number of vertices on one side of the axis of symmetry is not the same as the number of vertices on the other side of the axis of symmetry, which is impossible.

Example 2: Prove that the harmonic series $\sum_{n=1}^{\infty} \frac{1}{n}$ diverges.

This problem does not fit the "if … then" mold for choosing the proof by contradiction strategy. So what is there about this problem that would lead you to adopt this strategy? Primarily it is the fact that directly proving that the series diverges is hard, and it is often easier to prove something indirectly than to prove it directly. Adopting the proof by contradiction approach, you begin the proof by assuming that

the harmonic series $\sum_{n=1}^{\infty} \frac{1}{n}$ converges to some number k.

Assuming the series converges to k, you can write the following.

$$\sum_{n=1}^{\infty} \frac{1}{n} = \frac{1}{1} + \frac{1}{2} + \frac{1}{3} + \frac{1}{4} + \frac{1}{5} + \cdots = k$$

Can you use this equality to arrive at a contradiction? You can easily see that

$$\frac{1}{1} = \frac{1}{2} + \frac{1}{2}$$

$$\frac{1}{2} = \frac{1}{4} + \frac{1}{4}$$

$$\frac{1}{3} = \frac{1}{6} + \frac{1}{6}$$

$$\vdots$$

Using these equalities, you can see that k can be written as

$$k = \left(\frac{1}{2} + \frac{1}{2}\right) + \left(\frac{1}{4} + \frac{1}{4}\right) + \left(\frac{1}{6} + \frac{1}{6}\right) + \cdots$$

Now you can compare your original representation of k and the new one. Can you see the difference? Comparing terms of the two sums, you can see that $\frac{1}{1} > \frac{1}{2}$, $\frac{1}{3} > \frac{1}{4}$, $\frac{1}{5} > \frac{1}{6}$, and so on. You can easily see that the two representations of k are not equivalent, since

$$k = \frac{1}{1} + \frac{1}{2} + \frac{1}{3} + \frac{1}{4} + \frac{1}{5} + \frac{1}{6} + \cdots >$$

$$\frac{1}{2} + \frac{1}{2} + \frac{1}{4} + \frac{1}{4} + \frac{1}{6} + \frac{1}{6} + \cdots = k.$$

Since the inequality shows that $k > k$, you have reached a contradiction. Therefore, you can conclude that the given series diverges.

4. Working backward

Working backward is the strategy of starting at the conclusion to the problem and working, step by step, back to the information that was given in the problem statement. The solution to the problem is then to write the steps in order, from the information that was given to the desired conclusion. If you use this method, you must be sure that each step in the solution is reversible (i.e., if when "working backward," statement A is followed by statement B, then in the written solution, statement B must be followed by, and must imply, statement A).

Example 1: If a, b, and c denote the lengths of the sides of a triangle, show that

$$(a+b+c)^2 \leq 4(ab+bc+ca).$$

Why choose working backward to solve this problem? The information given is simple and does not, in any obvious way, contain anything related to the rather complicated formula in the conclusion. It is therefore useful to see whether you can simplify the inequality in the conclusion by using reversible steps.

This can be done as follows.

$$(a+b+c)^2 \leq 4(ab+bc+ca)$$

Either by recall or by multiplying out, you get

$$(a+b+c)^2 = a^2 + b^2 + c^2 + 2(ab+bc+ca)$$

Therefore, substituting for $(a+b+c)^2$ in the inequality, you get

$$a^2 + b^2 + c^2 + 2(ab+bc+ca) \leq 4(ab+bc+ca).$$

You can simplify the inequality even further. To do this, subtract $2(ab+bc+ca)$ from both sides of the inequality, and you will get the following.

$$a^2 + b^2 + c^2 \leq 2(ab+bc+ca)$$

By expanding and regrouping the right-hand side of the inequality, you will get a sum in the form a(something) $+ b$(something) $+ c$(something).

$$(a+b+c)^2 \leq a(b+c) + b(a+c) + c(b+a)$$

Now in order to compare the first term on the left-hand side to the first term on the right-hand side, and so on, it will be helpful to go back to the statement of the problem and recall that a, b, and c denote the lengths of the sides of a triangle. You know that the sum of any two sides of a triangle is larger than the remaining side (i.e., $a < b + c$, $b < a + c$, and $c < a + b$). Thus,

$$a^2 = a(a) \leq a(b+c)$$
$$b^2 = b(b) \leq b(a+c)$$
$$c^2 = c(c) \leq c(b+a)$$

and since the steps are reversible, your solution is complete.

Example 2: If n and m are integers and $i^n = -i$, show that $n = 4m - 1$.

As you can see, the information given is less complex than the conclusion, so working backward is a good strategy to try.

Assume that $i^{4m-1} = -i$. It will be helpful to recall that $i \times i = i^2 = -1$. Now, you can easily see that

$$i^{4m-1} = \left(i^{4m}\right)i^{-1}$$
$$= \left(i^4\right)^m i^{-1}$$
$$= \left(1^m\right)i^{-1}$$
$$= -i$$

Notice that all of the steps are reversible and i^4 is the first power of i that is equal to 1. Thus, by writing the steps in reverse for any integer m, you can see that

$$-i = \left(1^m\right)i^{-1}$$
$$= \left(i^4\right)^m i^{-1}$$
$$= i^{4m-1}$$

Hence, $n = 4m - 1$.

5. Using proof by induction

In elementary cases in which proof by induction is used, it is used to prove that a general statement about the positive integers is true.

To carry out this proof by induction, you need to

- verify that the proposed formula or theorem is true for $n = 1$, and
- assume that the proposed formula or theorem is true for $n = k$ (this is called the inductive hypothesis) and use this assumption to prove that it is true for $n = k + 1$. That is, if the proposition is true for any particular positive integer n, it must be true for the next larger positive integer value of n.

If you can carry out both these steps, you have proved that the general statement about the positive integers is true.

Proof by induction can be generalized to the case in which you are asked to show that a general statement about the sequence of integers $j, j+k, j+2k, \ldots, j+mk, \ldots$ where j and k are fixed positive integers, is true.

To carry out a proof by induction for this general case, you need to

- verify the proposed formula or theorem for the first integer j in the sequence of integers, and
- assume that the proposed formula or theorem is true for an integer i in the sequence, and use this assumption to prove that it is true for $i+k$. That is, if the proposition is true for any particular integer in the sequence, it must be true for the next integer in the sequence.

If you can carry out both these steps, you have proved that the general statement about the integers $j, j+k, j+2k, \ldots, j+mk, \ldots$ is true.

Example 1: Prove by induction or any other method that $n^3 - n$, where n is a positive integer, is divisible by 3.

Following the two steps given in the description of how to do proof by induction, we first need to show that the statement is true for $n=1$. Substituting 1 in the expression $n^3 - n$ we get $1^3 - 1 = 0$, which is divisible by 3, and thus the statement is true for $n=1$.

Next, following the second step in the description of how to do proof by induction, we assume that if $n=k$, then $k^3 - k$ is divisible by 3. We need to show that the statement is true for $n=k+1$. That is, we need to show that $(k+1)^3 - (k+1)$ is divisible by 3. Note that

$$
\begin{aligned}
&(k+1)^3 - (k+1) \\
&= \left(k^3 + 3k^2 + 3k + 1\right) - (k+1) \\
&= \left(k^3 - k\right) + 3\left(k^2 + k\right).
\end{aligned}
$$

You can see that both $k^3 - k$ and $3\left(k^2 + k\right)$ are divisible by 3 (the first by the assumption of the inductive hypothesis, and the second because it is 3 times some integer). Therefore, $(k+1)^3 - (k+1)$ is also divisible by 3. Thus we have shown that $n^3 - n$ is divisible by 3 for every positive integer n.

Example 2: If $n \geq 2$, prove that, in the xy-plane, the number of straight lines determined by n points, with no three points on the same straight line, is $\dfrac{n(n-1)}{2}$.

Following the two steps given in the description of how to do proof by induction, we first need to show that the statement is true for $n = 2$, the first integer in the set of integers for which we are proving the statement to be true. The statement is true when $n = 2$, since $\frac{2(2-1)}{2} = 1$ and two points determine one line.

Next, following the second step in the description of how to do proof by induction, we assume for $n = k$ that k points, with no three points on the same straight line, determine $\frac{k(k-1)}{2}$ lines. We need to show that the statement is true for $n = k + 1$. That is, we need to show that $k + 1$ points, with no three points on the same straight line, determine $\frac{k(k+1)}{2}$ lines.

The $k + 1$ points can be looked at as the addition of one point to an original collection of k points. By the inductive hypothesis, there are $\frac{k(k-1)}{2}$ lines determined by the original collection of k points. Since no three points lie on the same line, the additional point determines an additional k lines (each of these k lines is determined by the additional point and one of the k points in the original collection). Thus, altogether we have $\frac{k(k-1)}{2} + k = \frac{k(k+1)}{2}$ lines. Thus we have proved that if $n \geq 2$, the number of straight lines determined by n points, with no three points on the same straight line, is $\frac{n(n-1)}{2}$.

Thus, the general statement is true.

6. Solving a similar problem

If you notice that a problem is similar to one you have seen before, but the numbers or the words used to present the problem are different, try solving the problem the way you did before. It may be useful to try to modify the solution to the problem you already know how to solve to fit the new problem.

Example: Suppose you are asked to find a function $y = f(x)$ that is continuous but not differentiable at $x = 1$.

This problem is very similar to the problem of finding a function $y = f(x)$ that is continuous but not differentiable at $x = 0$. Recalling from examples you have seen that the function $y = |x|$ is continuous at $x = 0$ but not differentiable there, you can try to use this function to get to the desired function. Looking at the difference between the requirements for the two functions, you can see that the desired function can be arrived at by modifying the function $y = |x|$. You can easily see that this modification can be achieved by moving the entire graph of the function so that the point at which the moved function is not differentiable is $x = 1$. This is done by translating the graph of the function one unit to the right. Therefore, a function satisfying the desired conditions is $y = |x - 1|$.

7. Finding a pattern

Working with particular instances in which a general statement applies can help you get a feel for the problem and convince you of the plausibility of the result. When further exploration is done in a systematic way, patterns may emerge.

Patterns are found throughout mathematics. Identifying a pattern is often the first step in understanding a complex mathematical situation. Pattern recognition may yield insight that can point in the direction of a complete solution to the problem or simply help you generate a hypothesis about the problem situation, which you will need to explore further by using some other problem-solving strategy.

Example: Suppose that you are asked the following question.

If a positive integer k is a power of 3 (i.e., $k = 3^n$, where n is a positive integer), which of the ten digits 0, 1, 2, 3, 4, 5, 6, 7, 8, and 9 could be the units digit of k?

You can begin to approach this problem by looking at specific powers of 3 and seeing if a pattern emerges.

$$3^1 = 3$$
$$3^2 = 9$$
$$3^3 = 27$$
$$3^4 = 81$$
$$3^5 = 243$$
$$3^6 = 729$$
$$3^7 = 2,187$$

Notice that the units digits of the first seven powers of 3 are 3, 9, 7, 1, 3, 9, and 7. The obvious pattern is that the units digits of the powers of 3 cycle through the integers 3, 9, 7, and 1. This might lead you to suspect that the units digit of each power of 3 depends ONLY on the units digit of the previous power of 3. Thus, it seems likely that the answer to the question is that the units digit of a power of 3 must be 1, 3, 7, or 9. Given the pattern, a strategy for proving that your hypothesis is true is to perform the multiplications that yield the successive powers of 3 listed above to see whether the units digit of each power of 3 depends only on the units digit of the previous power of 3, and why.

8. Introducing the rectangular-coordinate system to solve geometry problems

When you are given a problem involving the properties of a geometric figure, it is often useful to introduce a *rectangular*-coordinate system and use the methods of analytic geometry to solve the problem.

Example: Suppose that you were asked to prove that if the diagonals of a parallelogram are congruent, the parallelogram is a rectangle.

For simplicity, you can put the parallelogram in the first quadrant of the xy-plane with one vertex at the origin and one side on the x-axis and make that side of length 1. This makes the coordinates of two of the vertices of the parallelogram $(0, 0)$ and $(1, 0)$. Giving the third vertex the coordinates (c, d), where c and d are greater than or equal to 0, you can use the properties of a parallelogram to conclude that the coordinates of the fourth vertex are $(1+c, d)$. The graph looks like this.

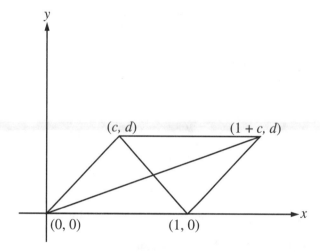

Now, you can recall that if two line segments are congruent, they have the same length. The lengths of the two diagonals are $\sqrt{(1+c)^2 + d^2}$ and $\sqrt{(c-1)^2 + d^2}$. Since the problem states that the diagonals of the parallelogram are congruent, you can conclude that the lengths are equal, so

$$\sqrt{(1+c)^2 + d^2} = \sqrt{(c-1)^2 + d^2}$$
$$(1+c)^2 + d^2 = (c-1)^2 + d^2$$
$$1 + 2c + c^2 + d^2 = c^2 - 2c + d^2 + 1$$
$$2c = -2c$$
$$c = -c$$
$$c = 0$$

Thus, the coordinates of the third and fourth vertices are $(0, d)$ and $(1, d)$ and the parallelogram is a rectangle.

9. Drawing a picture

For some problems, drawing a picture makes it easier for you to see relationships and dependencies that are not obvious when the information is presented in another way.

Example 1: A circle is circumscribed about a regular hexagon. If the area of a circular region is 2π, what is the area of the region of the regular hexagon?

If a geometry problem is not presented with an accompanying figure, you may want to start solving the problem by drawing a figure. It is a good idea for your figure to represent the given data accurately and for the unknowns to be labeled on the figure.

For this problem, you can draw the following figure.

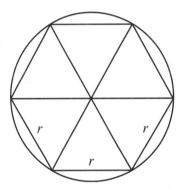

From the figure above, you can easily see that the hexagon can be broken down into six congruent equilateral triangles, each with sides of length r, where r is the radius of the circumscribed circle.

The area of each of the six equilateral triangles can be calculated using the formula $A = \frac{1}{2}bh$, where the base, b, of the triangle is equal to r, and the height, h, is the leg of a right triangle opposite a $60°$ angle. (The measures of the angles in an equilateral triangle are all equal to $60°$.) It follows that $h = r\sin(60°)$, and therefore $A = \frac{1}{2}r^2 \sin(60°)$. Thus, the area of the hexagonal region is $6\left(\frac{r^2 \sin(60°)}{2}\right) = 3r^2 \sin(60°)$.

You can then find the radius r by using the fact that the area of the circular region is given in the problem statement to be 2π. Therefore, $\pi r^2 = 2\pi$ and $r = \sqrt{2}$.

Since the $\sin(60°) = \frac{\sqrt{3}}{2}$, then the area of the regular hexagonal region is equal to

$$3 \cdot (\sqrt{2})^2 \sin 60° = 6\left(\frac{\sqrt{3}}{2}\right) = 3\sqrt{3}.$$

Example 2: Before conducting interviews, a certain company uses a preemployment test to screen all applicants. The test was passed by 60 percent of the applicants. Among those who passed the test, 80 percent were accepted for employment after being interviewed. Recently, the company interviewed a random sample of the applicants who did *not* pass the test. Of this group, 50 percent were accepted for employment. At these rates, what percent of applicants would be accepted for employment if no preemployment test were used?

Drawing figures or diagrams can be helpful, not only in geometric problems, but also in solving all sorts of problems in which there is a context that is not geometric. For example, for some probability and counting problems, using a diagram can make it easier for you to analyze the relevant data and to notice relationships and dependencies.

In this problem, applicants were sorted by means of a test and interview. It is hard to analyze what is happening to the applicant population using just words and numbers or symbols. A diagram can help clarify the situation. Therefore, you can start solving the problem by drawing the following diagram.

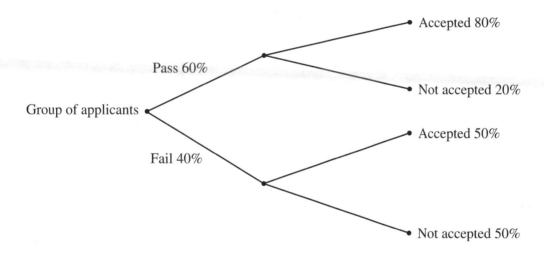

From this diagram, you can easily see that the percent of the applicants who passed the test, were interviewed, and accepted for employment is 80% of 60%, or 48%.

Similarly, the percent of the applicants who failed the test but were interviewed and accepted for employment is 50% of 40%, or 20%.

Again, from the diagram you can see that if no preemployment screen were used, the percent of applicants who would be accepted for employment would be the sum of the percents of those accepted for employment, which is 48% + 20%, or 68%.

Example 3: At how many points in the xy-plane does the graph of the function $y = x^2 - 4$ intersect the graph of the function $y = x^3 + x^2 - 2x + 4$?

(A) One
(B) Two
(C) Three
(D) Four

In problems involving the behavior of the graphs of functions in the xy-plane, it is often useful to graph the function using the graphing calculator and an appropriate viewing window.

Keeping in mind what the graphs of quadratic and cubic functions look like, you can see that using the viewing window $[-10, 10] \times [-10, 10]$ shows all points of intersection of the graphs of the two functions.

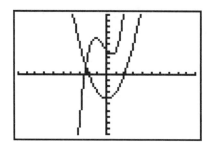

Therefore, the graphs of the functions intersect at one point, and the correct answer is A.

10. Adding lines to geometric figures

Adding "useful" lines to the figure is often a good strategy to use in solving problems involving geometric figures.

Example: Given triangle *ABC*, prove that if angle *A* is congruent to angle *C*, then triangle *ABC* is isosceles.

Notice that we are given information about triangle *ABC* without a picture of the triangle. Therefore, using the draw-a-picture strategy, our first step in solving the problem is to draw a triangle that meets the conditions in the problem statement.

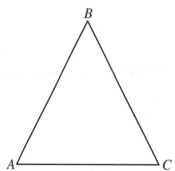

To prove that the triangle is isosceles, we need to show that *AB* is congruent to *BC*. The need to prove congruence suggests that it could prove useful to form two right triangles. We can do this by dropping a perpendicular from vertex *B* to side *AC*, as shown.

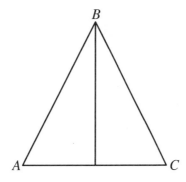

The rest of the proof can be completed using the properties of the right triangles formed by dropping the perpendicular.

11. Using symmetry

The presence of symmetry in a figure, diagram, or graph can help you see links and relationships that might be more difficult to discover by other means.

Example 1: In the *xy*-plane, which of the following is the equation of the graph produced by the reflection of the graph of $y = x^2 + 1$ over the *y*-axis?

(A) $y = x$

(B) $y = x^2$

(C) $y = x^2 - 1$

(D) $y = x^2 + 1$

Using the strategy of drawing the graph of a function in the *xy*-plane using a graphing calculator, you can see that the graph of $y = x^2 + 1$ is a parabola. Notice that the parabola is symmetric with respect to the *y*-axis.

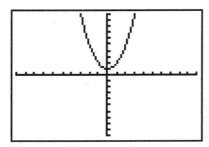

Recall that a reflection of the graph of an object over a line is simply a mirror image of the graph, where the mirror is placed on that line. Therefore, when you reflect a graph of $y = x^2 + 1$ over its axis of symmetry, the result is the same graph. Thus, the correct answer is D.

Example 2:

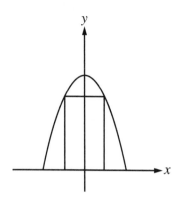

In the *xy*-plane above, a rectangle has two of its vertices on the *x*-axis and the other two on the graph of the parabola with equation $y = 16 - x^2$. Which of the following represents the area of the rectangular region?

(A) *xy*

(B) 2*xy*

(C) 3*xy*

(D) 4*xy*

You can see from the graph that the parabola is symmetric with respect to the *y*-axis (i.e., the same value of *y* results when *x* or −*x* is substituted into the equation $y = 16 - x^2$). Therefore, if one vertex of the rectangle has the coordinates (*x*, *y*), the corresponding vertex has coordinates (−*x*, *y*). Hence, you can conclude that the coordinates of the four vertices of the rectangle are (*x*, *y*), (−*x*, *y*), (*x*, 0), and (−*x*, 0), as indicated on the graph below.

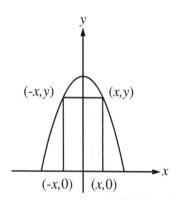

Now you can conclude that the dimensions of the rectangle are width 2*x* and length *y* and the area of the rectangular region is 2*xy*. Thus, the correct answer is B.

Example 3: The linear transformation T maps a point with coordinates (x, y) in 2-space into its reflection over the x-axis. Which of the following matrices represents the inverse of T?

(A) $\begin{pmatrix} -1 & 0 \\ 0 & -1 \end{pmatrix}$

(B) $\begin{pmatrix} -1 & 0 \\ 0 & 1 \end{pmatrix}$

(C) $\begin{pmatrix} 1 & 0 \\ 0 & -1 \end{pmatrix}$

(D) $\begin{pmatrix} 1 & 0 \\ 0 & 1 \end{pmatrix}$

The figure below might help you grasp the problem better.

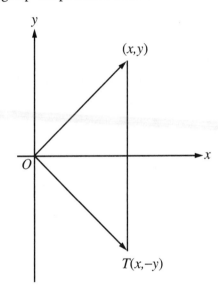

From the figure, you can see that the linear transformation T sends (x, y) to $(x, -y)$. You can find the matrix associated with this transformation by describing its effects on the basis vectors $(1, 0)$ and $(0, 1)$. Thus, $T(1, 0) = (1, 0)$ and $T(0, 1) = (0, -1)$.

Hence,

$$T(x, y) = \begin{pmatrix} 1 & 0 \\ 0 & -1 \end{pmatrix}\begin{pmatrix} x \\ y \end{pmatrix}$$

Now is a good time to reexamine the figure. You can see that the figure is symmetric with respect to the x-axis. What does this tell you? One thing it tells you is that the inverse of T will be given by reflection over the x-axis (i.e., the transformation is its own inverse). Since the inverse of T is T, the matrix representing the inverse of T is the matrix representing T, or $\begin{pmatrix} 1 & 0 \\ 0 & -1 \end{pmatrix}$. Thus, the correct answer is C.

12. Solving a simpler related problem

What happens if you cannot see clearly how various pieces of information given in a problem are connected? This could happen because the relationships between the pieces of information are very complicated. It is sometimes helpful to try to solve a simpler related problem.

Example 1: If 7 different points on a circle are selected, what is the number of quadrilaterals that have 4 of these points as vertices?

(A) 35
(B) 49
(C) 210
(D) 840

The relationship between the 7 points on the circle and the number of quadrilaterals is complicated. It is reasonable to try to solve the simpler, related problem, "If 4 different points are selected on a circle, what is the number of quadrilaterals that have these points as vertices?" This simpler problem can be approached by using the strategy of drawing a picture, as shown below.

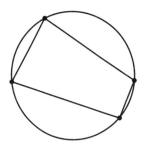

Notice that the four different points on the circle determine a unique quadrilateral. If you now look at the original problem, you can see that different selections of 4 points from the 7 on the circle will determine different quadrilaterals. There are $\binom{7}{4} = \frac{7!}{(7-4)!(4!)} = 35$ ways of choosing 4 points from a set of 7 points. Thus, the number of quadrilaterals is 35.

Example 2: A convex polygon is a polygon in which the measure of each interior angle is less than 180°. Which of the following could be the number of diagonals of a convex polygon?

I. 27
II. 45
III. 44

(A) I only
(B) II only
(C) I and III only
(D) I, II, and III

After looking at this example, you are probably asking yourself, "Do I have to examine all possible convex polygons?" The idea of looking at several polygons with a large number of sides could be overwhelming. If you are facing this situation, think how you can make the problem more manageable. You can do this by first looking at a rectangle. Each vertex of the rectangle has a diagonal connecting it to $4 - 3 = 1$ other vertex (i.e., to all vertices except itself and the two adjacent vertices). Now try to consider all 4 vertices. You can easily see that each diagonal will be encountered twice, therefore the number of diagonals is

$$\frac{\text{(number of vertices)} \times \text{(number of diagonals connecting each vertex to other vertices)}}{2} = \frac{4(4-3)}{2} = 2$$

Now you are ready to consider an *n*-sided convex polygon. Try to apply the same line of reasoning in this case that you used for the rectangle. Each vertex of the *n*-sided polygon has a diagonal connecting it to $n - 3$ other vertices (i.e., to all vertices except itself and the two adjacent vertices). Considering all *n* vertices and noting that each diagonal will be counted twice, you will find that the number of diagonals is equal to

$$\frac{\text{(number of vertices)} \times \text{(number of diagonals connecting each vertex to other vertices)}}{2} = \frac{n(n-3)}{2}$$

Now you can draw the following table.

Number of sides of a convex polygon (*n*)	9	10	11
Number of diagonals	27	35	44

Thus, the correct answer is C.

13. Dividing a problem into cases

Another effective strategy for showing that a general statement is true is to simplify the problem by dividing it into a number of cases, each of which you can handle separately. Every instance of the general statement must be covered by one of the cases, so that taking the cases together, you will have proved that the general statement is true. For example, if you want to prove that a statement is true for all integers, you may do this by proving that the statement is true for all even integers and all odd integers, since every integer is either even or odd.

If you decide to approach a problem this way, you will need to decide how to determine the cases to use. The cases you decide upon should be chosen so that they cover all instances of the general statement and simplify the analysis you have to do to solve the problem.

Example: If f is a function on the integers satisfying $f(x+y) = f(x) + f(y)$ for all integers x and y, prove that $f(x) = xf(1)$.

Looking at the information in the problem, you are given a relationship between $f(x)$ and $f(1)$ as well as a relationship between function values involving addition. The positive integers are first "built" by addition from 1. It is reasonable, therefore, to break the problem into two cases—the positive integers and the nonpositive integers.

Case I: For the positive integers:

> The result holds when $x = 1$, since $f(1) = 1f(1)$.
> For $x = 2$, you have $f(2) = f(1+1) = f(1) + f(1) = 2f(1)$.
> For $x = 3$, you have $f(3) = f(2+1) = f(2) + f(1) = 2f(1) + f(1) = 3f(1)$.
>
> Now you can see that this process can be continued, and you can use the pattern to develop a proof by induction that $f(n) = nf(1)$ for any positive integer n.

Case II: For the non-positive integers:

> In this case, you want to begin with 0 and count downward.
> If $x = 0$, you can see that $f(0) = f(0+0) = f(0) + f(0)$. Subtracting $f(0)$ from each side, you get $f(0) = 0$. That is, $f(0) = 0 \times f(1)$.
>
> You can use the function values for the nonnegative integers to get the value of $f(-1)$. Using the equality $0 = 1 + (-1)$, you can write $0 = f(0) = f(1 + (-1)) = f(1) + f(-1)$.
>
> From this you can see that $f(-1) = -f(1)$. Similarly, $0 = f(0) = f(n + (-n)) = f(n) + f(-n)$. From this you can see that $f(-n) = -f(n)$. Using the identity that you proved for the positive integers, that is, $f(n) = nf(1)$ for any positive integer n, you can easily see that $f(-n) = -nf(1)$ for any positive integer n.

14. Looking for another way to solve the problem

Sometimes you may face a situation in which you read and understand the problem and choose a strategy, but still can't get a solution. Don't get discouraged! Step back for a moment and think. Remember that a lot of problems in mathematics have more than one solution. If your chosen strategy is not working, don't get bogged down; look for another way to solve the problem.

Example 1: $0 < x < \dfrac{\pi}{2}$ and $\tan x = \dfrac{3}{5}$, what is the value of $\cos x$?

Solution 1: When you are solving problems in trigonometry, it is always helpful to recall some trigonometric identities. Those trigonometric formulas might significantly simplify the solution. Recall that

$$\sec^2 x = \frac{1}{\cos^2 x} \quad \text{and} \quad \sec^2 x = 1 + \tan^2 x$$

Since $\sec^2 x = 1 + \tan^2 x$ and $\tan x = \dfrac{3}{5}$, then $\sec^2 x = 1 + \dfrac{9}{25} = \dfrac{34}{25}$.

Therefore, $\cos^2 x = \dfrac{25}{34}$.

Since $0 < x < \dfrac{\pi}{2}$ requires $\cos x > 0$, you will find that $\cos x = \sqrt{\dfrac{25}{34}} = \dfrac{5\sqrt{34}}{34}$.

Solution 2: If you do not see how to use these identities, consider possible geometric solutions.

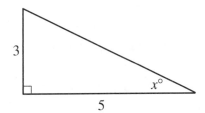

In the triangle above, $\tan x = \dfrac{3}{5}$. The hypotenuse of this triangle is equal to

$\sqrt{3^2 + 5^2} = \sqrt{34}$, and $\cos x = \dfrac{5}{\sqrt{34}} = \dfrac{5\sqrt{34}}{34}$.

Example 2: $\lim\limits_{x \to 0} \dfrac{(\sin 2x)(\tan x)}{3x}$ is

(A) 0

(B) $\dfrac{1}{3}$

(C) $\dfrac{2}{3}$

(D) nonexistent

Solution 1: You can see that the function $f(x) = \dfrac{(\sin 2x)(\tan x)}{3x}$ takes the indeterminate form $\dfrac{0}{0}$ when $x \to 0$, and therefore you can apply L'Hôpital's rule. Differentiating numerator and denominator, you get

$$\lim_{x \to 0} \frac{(\sin 2x)\left(\sec^2 x\right) + (2\cos 2x)(\tan x)}{3} = 0$$

Thus, the correct answer is A.

Solution 2: If you recognize that for all small angles θ, $\sin\theta \approx \theta$ and $\tan\theta \approx \theta$, the given expression becomes

$$\lim_{x \to 0} \frac{2x(x)}{3x} \approx \lim_{x \to 0} \frac{2x}{3} = 0.$$

15. Using a graphing calculator

Sometimes the strategies that you choose to tackle a problem are really too hard or too time-consuming to implement by hand. In such cases, try to use a graphing calculator. The graphing calculator can shorten the time it takes to perform computations, produce a graph, or do other mathematical procedures. However, keep in mind that the results provided by a graphing calculator supplement, but do not replace, your knowledge of mathematics. You must use your knowledge of mathematics to determine whether the information provided by the calculator is correct and complete. The graphing calculator is a very powerful tool, but it has limitations. The following is a discussion of some things to watch out for when you use the graphing capabilities of the graphing calculator.

Limitation 1: The resolution of the graph depends on the scale on the *x*- and *y*-axes that you choose. The graphing calculator produces a graph by plotting a large, but finite, number of points on the graph. The behavior of the function between the plotted points (pixels) is not shown. If the behavior of the function between the pixels is "nice," then the graphing calculator's graph "tells the truth" about the function. If the behavior of the function between the pixels is "not nice," then the graphing calculator's graph "lies." When you are working with a particular function using the graphing calculator, it is your knowledge of mathematics that enables you to determine whether the calculator is "telling the truth" or "lying."

Example 1: Try to plot the graph of the very familiar function $y = \sin x$ with the viewing window $[0, 64\pi] \times [-1, 1]$. The calculator will give you this graph.

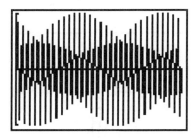

It surely does not look like a graph of $y = \sin x$.

Example 2: Try to plot the graph of the function $y = \dfrac{1}{\sqrt{x}}$ with the viewing window $[-10, 10] \times [-10, 10]$. The calculator will give you this graph.

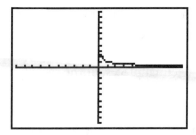

Your mathematical knowledge tells you that this function diverges to infinity as x approaches 0 from the right, but the graph does not tell you that visually.

Limitation 2: The graph of the function displayed on the screen of the graphing calculator shows only a portion of the entire graph of the function. The behavior of the function outside the values included in the viewing window is not shown. If the behavior of the function outside the viewing window is unimportant to the analysis of the function, then the graphing calculator's graph "tells the truth" about the function. If the behavior of the function outside the viewing window is important to the analysis of the function, then the graph "lies." When you are working with a particular function using the graphing calculator, it is your knowledge of mathematics that enables you to determine whether the calculator is "telling the truth" or "lying."

Example 1: At how many points in the xy-plane does the graph of the function $y = x^4 + 14x^3 + 23x^2 - 14x - 24$ intersect the x-axis?

$\frac{14}{\frac{3}{2}}$

(A) One
(B) Two
(C) Three
(D) Four

$y' = 4x^3 + 42x^2 + 46x - 14$

3 switches

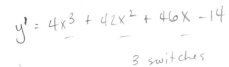

To find the number of points at which the graph of the given function intersects the x-axis, it is helpful to graph the function on a graphing calculator. With the viewing window $[-10, 10] \times [-10, 10]$, the calculator gives you this graph.

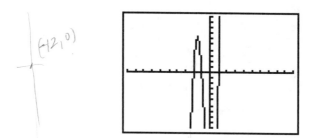

If you accept this result as the answer to the problem, you would say that the graph intersects the x-axis three times. Notice, however, that the function is a fourth-degree polynomial and the graph produced by the graphing calculator looks like the graph of a third-degree polynomial. Because of this, you know that not all of the intersection points of the function may be shown in the viewing window. Adjusting the viewing window to give you a graph that looks like a fourth-degree polynomial allows you to see that the function intersects the x-axis four times. Hence, the correct answer is D.

Example 2: What is the area, to two decimal places, of the region in the first quadrant of the xy-plane that is bounded above by the curve $y = 3 \sin x$ and below by the line $y = x$?

You can begin by graphing the functions $y = 3 \sin x$ and $y = x$ on a graphing calculator. With the viewing window $[-2\pi, 2\pi] \times [-4, 4]$, the calculator gives you this graph:

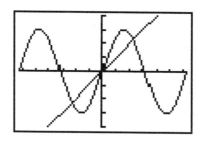

Using your knowledge of mathematics, you can conclude that neither limitation 1 nor limitation 2 applies in this case. You can represent the area of the desired region as $\int_0^b (3\sin x - x)\, dx$, where b is the x-coordinate of the point of intersection in the first quadrant of the two graphs. You can use the graphing calculator to determine that $b \approx 2.279$. Therefore, again using the graphing calculator, the area of the desired region is $\int_0^{2.279} (3\sin x - x)\, dx = \int_0^{2.279} (3\sin x)\, dx - \int_0^{2.279} x\, dx \approx 4.951 - 2.597 \approx 2.35$. Note that there are different ways of calculating both b and the integral with the graphing calculator and that different ways of doing the calculations will give you slightly different answers.

Chapter 5

Succeeding on Multiple-Choice and Constructed-Response Questions

► ► ► ► ► ► ► ► ► ► ► ►

$Int @ x = 2.279$

$\int_0^{2.279} 3\sin x = 4.951$

$\int_0^{2.279} x = 2.597$

2.279

2.279

2.279

$= 2.354 \checkmark$

0

The goal of this chapter is to provide you with background information and advice from experts so that you can improve your skill in answering multiple-choice questions and writing answers to constructed-response questions about mathematics.

Tools: Calculators and Commonly Used Formulas

It's important to remember the tools that will be available to you during the test administration. As you prepare for one or more of the mathematics tests, it's probably a good idea to use these tools as you would at an actual administration.

Calculators

For the *Mathematics: Content Knowledge* and the *Mathematics: Proofs, Models, and Problems, Part 1* tests, you will be expected to use a graphing calculator without a QWERTY (typewriter-layout) keyboard. Calculator memories need NOT be cleared. Your calculator should be able to

- Produce the graph of a function within an arbitrary viewing window
- Find the zeros of a function
- Compute the derivative of a function numerically
- Compute definite integrals numerically

As you take the practice tests and check your answers in this study guide, use your calculator to answer questions when appropriate. (See strategies later in this chapter for explaining calculator-derived results in your constructed-response answers.)

Commonly used formulas

You will notice that the practice tests contain four pages of commonly used notations, formulas, and definitions. These pages will be exactly the same in the test book at your actual administration. Get to know what is contained in these four pages so that during the actual test you will not have to use valuable time becoming familiar with the information contained there. As you take the test, you can refer to these pages as needed to find information to help you answer the questions.

Multiple-Choice Questions

Multiple-choice questions in mathematics are different from those in most other subject areas in several important ways.

Wording

Mathematics questions are economically worded—that is, every word counts in terms of information you need to get to the right answer. Therefore, it pays to read the questions slowly and carefully to identify and understand each piece of information that is needed to find a solution. Here is an example of a typical mathematics question.

If (i) the graph of the function $f(x)$ is the line with slope 3 and y-intercept 1 and

   (ii) the graph of the function $g(x)$ is the semicircle in the upper half plane with center at the origin and radius 2,

what is the domain of $g(f(x))$?

(A) [0, 2]

(B) $\left[-1, \dfrac{1}{3}\right]$

(C) [−2, 2]

(D) (−∞, ∞)

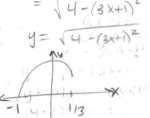

The structure of the question is simple and concise: "If" a and b, then "what?" Given two pieces of information, what is the third piece of information?

It is critical to look carefully at each of the three parts of the question. Within each statement there is a wealth of information. The first statement reads:

  (i) the graph of the function $f(x)$ is the line with slope 3 and y-intercept 1

From this statement, you need to figure out that the graph of the function $f(x)$ is the line represented by $y = 3x + 1$.

The second statement provides another critical piece of information:

  (ii) the graph of the function $g(x)$ is the semicircle in the upper half plane with center at the origin and radius 2

From this statement, you need to figure out that the function $g(x)$ is $\sqrt{4 - x^2}$. Here is how to do this. Recall that the equation of a circle with center at (0, 0) and radius 2 can be written as $x^2 + y^2 = 4$.

Using some elementary algebra, you can rewrite this as $y^2 = 4 - x^2$ or $y = \pm\sqrt{4 - x^2}$. (Remember that this is not a function.) The graph of $g(x)$, the semicircle in the upper half of the plane, is represented by $y = \sqrt{4 - x^2}$. By sketching and inspecting the graph of $g(x)$, you can infer that the domain of $g(x)$ is $-2 \le x \le 2$. You can confirm this by graphing the semicircle as well as the line on a calculator using the viewing window $-4 \le x \le 4$ and $-3 \le y \le 3$.

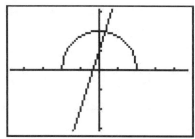

The third part of the question presents the way you are to put the first two parts together:

> what is the domain of $g(f(x))$?

You need to understand from this part of the question that the domain of $g(f(x))$ is the set of all x for which $g(f(x))$ is defined.

Since $g(f(x))$ is defined for all x such that $-2 \le f(x) \le 2$, and since $f(x)$ is given by the equation $f(x) = 3x + 1$, the domain of $g(f(x))$ is the set of all x satisfying the double inequality

$-2 \le 3x + 1 \le 2$

$-3 \le 3x \le 1$

$-1 \le x \le \dfrac{1}{3}$

[Handwritten annotation: Domain $g(x) \Rightarrow -2 \le x \le 2$; Domain $g(f(x)) \Rightarrow -2 \le f(x) \le 2$; $-2 \le 3x+1 \le 2$]

It follows that the domain of $g(f(x))$ is the closed interval $\left[-1, \dfrac{1}{3}\right]$; thus the correct answer is B.

You can solve the problem a different way. You can let $Y_1 = 3x + 1$, $Y_2 = \sqrt{4 - x^2}$, and $Y_3 = Y_2(Y_1(x))$, or $Y_3 = \sqrt{4 - (3x + 1)^2}$.

The domain of Y_3 is those values of x for which $4 - (3x + 1)^2 \ge 0$. By inspection and comparison with the inequality above, you will find that $-1 \le x \le \dfrac{1}{3}$.

This can be confirmed by graphing Y_3 on your calculator and inspecting the graph, which shows x to have values between -1 and $\dfrac{1}{3}$.

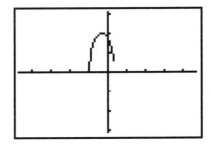

B is the only answer choice consistent with this graph; therefore the correct answer is B.

As this problem shows, the length of the question itself does not necessarily indicate how many steps will be needed to solve it, how a graphing calculator should be used, or how difficult or easy the problem is. In multiple-choice questions in mathematics, you must take your time, read carefully, and use every bit of information from each word, phrase, and sentence.

Timing

Multiple-choice questions in mathematics are also different from questions in many other subject areas, because they can vary greatly in the time it takes to arrive at the correct answer. Some of the problems are more conceptual, are relatively straightforward, and take little time to solve, provided you know the concept. Others require more effort to read, understand, set up, and solve, even when you know your content extremely well. It is fine to spend a little extra time trying to solve the longer problems, provided you can make it up on other problems. For the *Content Knowledge* test, you have two hours to complete 50 questions, which means almost two-and-a-half ($2\frac{1}{2}$) minutes per question. You should expect some questions to take more time than that, and some to take less. During the test, you may want to consider checking your progress in half-hour increments—you should complete at least 12 or 13 questions each half-hour in order to finish the test in two hours.

Multiple ways of finding a solution

There is often more than one way of arriving at a correct answer in a mathematics question, and you should rely on your own ways of solving problems when you take the test. If you tend to draw pictures or make graphs, then use the blank spaces in the test book to do so. If you typically use your calculator for virtually all problem solving, then use it in that way during the test. Whatever ways you use to work math problems, however, be sure to read the question carefully, use every bit of information given, and carefully but efficiently set up the path to the correct solution.

Multiple-Choice Tips

Here are some additional expert tips to help you succeed on the mathematics multiple-choice questions.

Look at your choices before solving. Scan the answer choices offered before you start working on a problem. This might give you an idea of the sort of answer you should be looking for and may also give you a clue about how to solve the problem.

Use a process of elimination. In a multiple-choice question, you know that of the four answer choices, one is correct and three are incorrect. If you eliminate three of the choices as possible answers, then the choice that you have not eliminated must be the correct answer.

Use estimation. Sometimes you can save time and effort by simply estimating the correct answer. After that, you can scan the answer choices and pick the one closest to your estimate.

Use good judgment when checking a calculator-derived answer. You may find that the answer you produce with the help of the calculator does not match any of the answer choices exactly. This can occur because calculator output will vary slightly depending on the method by which the output is produced. In this case, scan the choices. One of them should differ only slightly from your answer. If none of the answer choices is close to your answer, look for an error in how you solved the problem.

Constructed-Response Questions

What you should know about how the *Mathematics* Constructed-Response Tests are scored

As you build your skills in writing answers to constructed-response questions, you should keep in mind the process used to score the tests. If you understand the process by which experts award your scores, you may have a better context in which to think about your strategies for success.

How the tests are scored

After each test administration, test books are returned to Educational Testing Service (ETS.) The test booklets in which constructed-response answers are written are sent to the location of the scoring session.

The scoring session usually takes place over two days. It is led by scoring leaders, highly qualified mathematics educators who have many years' experience scoring test questions. All of the remaining scorers are experienced mathematics teachers and educators of mathematics teachers. ETS attempts to balance experienced scorers with newer scorers at each session; the experienced scorers provide continuity with past sessions, and the new scorers ensure that new perspectives are incorporated and that the pool of scorers remains large enough to cover the test's needs throughout the year.

Preparing to train the scorers

The scoring leaders meet several days before the scoring session to assemble the materials for the training portion of the main session. The process of training scorers is a rigorous one, and it is designed to ensure that each response receives a score that is consistent both with the scores given to other papers and with the overall scoring philosophy and criteria established for the test when it was designed.

The scoring leaders first review the General Scoring Guides, which contain the overall criteria, stated in general terms, for awarding the appropriate score. The leaders also review and discuss—and make additions to, if necessary—the Question-Specific Scoring Guides, which serve as applications of the general guide to each specific question on the test. The question-specific guides cannot cover every possible response the scorers will see, but they are designed to give enough examples to guide the scorers in making accurate judgments about the variety of answers they will encounter.

To begin identifying appropriate training materials for an individual question, the scoring leaders first read through many responses to get a sense of the range of the answers. They then choose a set of benchmarks, selecting one paper at each score level. These benchmarks serve as solid representative examples of the kind of response that meets the criteria of each score level and are considered the foundation for score standards throughout the session.

The scoring leaders then choose a larger set of test-taker responses to serve as sample papers. These sample papers represent the wide variety of possible responses that the scorers might see. The sample papers serve as the basis for practice scoring at the scoring session, so that the scorers can rehearse how they will apply the scoring criteria before they begin.

The process of choosing a set of benchmark responses and a set of sample responses is followed systematically for each question to be scored at the session. After the scoring leaders are done with their selections and discussions, the sets they have chosen are photocopied and inserted into the scorers' folders in preparation for the main scoring session.

Training at the main scoring session

At the scoring session, the scorers are placed into groups according to the question they are assigned to score. New scorers are distributed equally across these groups. One of the scoring leaders is placed with each group. The Chief Scorer is the person who has overall authority over the scoring session and plays a variety of key roles in training and in ensuring consistent and fair scores.

For each question, the training session proceeds in the same way:

1. All scorers carefully read through the question they will be scoring.

2. All scorers review the General Scoring Guide and the Question-Specific Scoring Guide for the question.

3. For each question, the leader first guides the scorers through the set of benchmark responses, explaining in detail why each response merited the score it did. Scorers are encouraged to ask questions and share their perspectives.

4. Scorers then practice on the set of sample responses. The leader polls the scorers on what scores they would award and then leads a discussion to ensure that there is consensus about the scoring criteria and how they are to be applied.

5. Each member of the group then reads one or more sets of unscored papers (five to eight papers in a set). After each set is read, the scores awarded and the scoring criteria are discussed. The papers are then scored using a consensus scoring technique.

6. When the leader is confident that the scorers for that question will apply the criteria consistently and accurately, the actual scoring begins.

Quality-control processes

A number of procedures are followed to ensure that accuracy of scoring is maintained during the scoring session. Most importantly, each response is scored twice, with the first scorer's decision hidden from the second scorer. If the two scores for a paper are the same or differ by only one point, the scoring for that paper is considered complete, and the response is awarded the sum of the two scores. If the two scores differ by more than one point, the response is scored by a scoring leader, who is not permitted to see the two other scorers' decisions. If this third score is midway between the first two scores, the score for the question is the sum of the first two ratings (e.g., 5, 3, and 4 = final score of 8); otherwise it is the sum of the third score and whichever of the first two scores is closer to it (e.g., 5, 3, and 3 = final score of 6).

Another way of maintaining scoring accuracy is through back-reading. Throughout the session, the leader for each question checks random samples of scores awarded by all the scorers. If the leader finds that a scorer is not applying the scoring criteria appropriately, that scorer is given more training.

At the beginning of the second day of each reading, additional sets of papers are scored using the consensus method described above. This helps ensure that the scorers are refreshed on the scoring criteria and are applying them consistently.

Finally, the scoring session is designed so that several different readers (at least four) contribute to any single test taker's score. This minimizes the effects of a scorer who might score slightly more rigorously or generously than other scorers.

The entire scoring process—general and specific scoring guides, standardized benchmarks and samples, consensus scoring, adjudication procedures, back-reading, and rotation of tests to a variety of scorers—is applied consistently and systematically at every scoring session to ensure comparable scores for each administration and across all administrations of the test.

General scoring guides

It is critical to understand the scoring guide that the scorers will be using when they score your test. If you familiarize yourself with what is important to the scorers, you should be able to maximize your chances for a successful score.

All of the questions on the mathematics constructed-response tests are scored holistically (that is, as a whole rather than in parts). The score range is 0 to 5. The two tests (*Proofs, Models, and Problems, Part 1*, and *Pedagogy*) have different scoring guides, since the tests assess different sets of knowledge and skills. Take some time as you read these guides to translate what the scorers are looking for into your own words, because you want the characteristics of the high end of the scale to become characteristics of your answers.

Mathematics: Proofs, Models, and Problems, Part 1

Each of the four questions is given a single score, ranging from 0 to 5.

Score	Comment
5	■ Clearly demonstrates a full understanding of the mathematical content necessary to answer all parts of the question successfully
	■ Gives a correct and complete response but may contain a minor calculation error
4	■ Clearly demonstrates a full understanding of the mathematical content necessary to answer all parts of the question successfully
	■ EITHER gives a correct and complete response that contains a minor calculation error or misstatement OR gives a correct and almost complete response
3	For a one-part question
	■ Clearly demonstrates an understanding of all aspects of the question
	■ Demonstrates the ability to determine an appropriate strategy for answering the question
	■ Makes substantial progress toward a correct and complete response
	For a multipart question
	■ Clearly demonstrates a full understanding of the mathematical content needed to answer a significant portion of the question successfully
	■ Gives a correct and complete response to that portion of the question

Score	Comment

2   For a one-part question

- EITHER demonstrates a limited understanding of the question OR makes only minimal progress toward a correct and complete response

For a multipart question

- Clearly demonstrates a full understanding of the mathematical content needed to answer a minor portion of the question successfully

- Gives a correct and complete response to that portion of the question

1

- Demonstrates a very limited understanding of the question or questions

- Makes little or no progress toward a correct and complete response

0

- Blank, almost blank, or off topic

Mathematics: Pedagogy

Each of the three questions is given a single score, ranging from 0 to 5.

A response that does not demonstrate an understanding of the mathematics to be presented CANNOT receive a score above 2, regardless of any other criteria for higher scores it might meet.

Score	Comment

5

- Clearly demonstrates an understanding of the mathematics to be presented

- Clearly explains how to present the mathematics to students in a way that is likely to achieve the desired goal(s)

- Gives a clear and complete response, develops the mathematics in a way that is well motivated (that is, students can clearly see why the mathematics being presented is worth studying and/or can see the mathematics as the logical consequence of previously studied mathematics)

4

- Clearly demonstrates an understanding of the mathematics to be presented, but may have a notational error or minor mathematical misstatement

- Explains how to present the mathematics to students in a way that can reasonably be expected to achieve the desired goal(s)

- EITHER gives an almost complete response and a well-motivated development of the material OR gives a complete response and a fairly well motivated development of the material

3   ■ Demonstrates an understanding of the mathematics to be presented

   ■ Indicates how to present the mathematics to students in a way that can reasonably be expected to achieve the desired goal(s)

   ■ EITHER gives an almost complete response and a well-motivated development of the material, OR gives a complete response and a fairly well motivated development of the material

2   ■ EITHER demonstrates a limited understanding of the mathematics to be presented (and may or may not indicate how to teach the mathematics to students in a way that is likely to achieve the desired goal[s]), OR demonstrates an understanding of the mathematics, but gives little or no indication of how to present the mathematics to students in a way that is likely to achieve the desired goal(s)

   ■ Gives an unclear and incomplete response

1   ■ EITHER demonstrates a very limited understanding of the mathematics to be presented, OR fails to discuss the mathematics at all

0   ■ Blank, almost blank, or off topic

Advice from the experts

Scorers who have scored thousands of real tests indicate the following practical pieces of advice.

Justify and explain. You will be scored on your ability to explain the mathematics involved in solving the questions, not just the ability to arrive at a correct answer. Your response should not contain statements without written justifications, particularly when you are producing a proof or answering a question that asks you to explain your reasoning. Similarly, do not write something such as, "This is the graph that appears on my calculator," in a response without a mathematical reason for why you have accepted the graph as correct and complete.

Show step-by-step work. If a question asks you to show your work or show how you arrived at your answer, your response can be in words or a series of mathematical steps that someone looking at your response can follow to see what mathematical steps you used to arrive at your answer.

Answer all parts. You must answer all parts of a question—and you must show your work on every part. The scorers must see that you have responded clearly and thoroughly to each part of each question. Be sure to review your responses before you hand in your test.

Write clearly and carefully. This will make it easier for you to check your work for obvious errors, such as a missing minus sign. Also, this will make it easier for the scorers to read and understand your responses. You will not be judged on your handwriting, but your response should be as clear as you can make it.

Make things crystal clear. When you are using variables to model a real-life application, state what each variable represents. In a problem involving a graph or a figure, label all parts of the graph or figure. If you use a graph produced by the graphing calculator, indicate what the viewing window is, or, in some other way, put a scale on the x- and y-axes.

Build on your solutions in multi-part questions. Constructed-response problems often consist of two or three different parts. One way to look at those parts is to consider them chains of related questions. One question will help you answer the next one. You can guess the answers by experimenting with the first few questions in the chain. However, after you are sure the answers are correct, *do not forget to justify each part*.

Chapter 6
Practice Test, *Mathematics: Content Knowledge*

▶   ▶   ▶   ▶   ▶   ▶   ▶   ▶   ▶   ▶   ▶   ▶

Practice Questions

Now that you have studied the content topics and have worked through strategies relating to the *Content Knowledge* test, you should take the following practice test. You will probably find it helpful to simulate actual testing conditions, giving yourself 120 minutes to work on the questions. You can use the answer sheet provided if you wish.

Keep in mind that the test you take at an actual administration will have different questions. You should not expect your level of performance to be exactly the same as when you take the test at an actual administration, since numerous factors affect a person's performance in any given testing situation.

When you have finished the practice questions, you can score your test and read the explanations of right answers in chapter 9.

THE PRAXIS™
S E R I E S

TEST NAME
Mathematics: **Content Knowledge (0061)**

Time—120 Minutes

50 Questions

DO NOT USE INK

Use only a pencil with soft black lead (No. 2 or HB) to complete this answer sheet.
Be sure to fill in completely the oval that corresponds to the proper letter or number.
Completely erase any errors or stray marks.

THE PRAXIS SERIES®
Professional Assessments for Beginning Teachers®

Answer Sheet C   PAGE 1

1. NAME
Enter your last name and first initial.
Omit spaces, hyphens, apostrophes, etc.

Last Name (first 6 letters)   F I

2.

YOUR NAME: (Print)
Last Name (Family or Surname)   First Name (Given)   M. I.

MAILING ADDRESS: (Print)
P.O. Box or Street Address   Apt. # (if any)

City   State or Province

Country   Zip or Postal Code

TELEPHONE NUMBER: (   )   Home   (   )   Business

SIGNATURE: _____   **TEST DATE:** _____

3. DATE OF BIRTH
Month   Day
Jan., Feb., Mar., April, May, June, July, Aug., Sept., Oct., Nov., Dec.

4. SOCIAL SECURITY NUMBER

5. CANDIDATE ID NUMBER

6. TEST CENTER / REPORTING LOCATION
Center Number   Room Number
Center Name
City   State or Province
Country

7. TEST CODE / FORM CODE

8. TEST BOOK SERIAL NUMBER

9. TEST FORM

10. TEST NAME

51055 • 08920 • TF71M500   Q2573-06
MH01159

I.N. 202974

Educational Testing Service, ETS, the ETS logo, and THE PRAXIS SERIES:PROFESSIONAL
ASSESSMENTS FOR BEGINNING TEACHERS and its logo are registered trademarks of
Educational Testing Service.

ETS Educational Testing Service

1 2 3 4

CERTIFICATION STATEMENT: (Please write the following statement below. DO NOT PRINT.)

"I hereby agree to the conditions set forth in the *Registration Bulletin* and certify that I am the person whose name and address appear on this answer sheet."

SIGNATURE: _____

DATE: _____ / _____ / _____
Month   Day   Year

BE SURE EACH MARK IS DARK AND COMPLETELY FILLS THE INTENDED SPACE AS ILLUSTRATED HERE: ● .

1 Ⓐ Ⓑ Ⓒ Ⓓ	41 Ⓐ Ⓑ Ⓒ Ⓓ	81 Ⓐ Ⓑ Ⓒ Ⓓ	121 Ⓐ Ⓑ Ⓒ Ⓓ
2 Ⓐ Ⓑ Ⓒ Ⓓ	42 Ⓐ Ⓑ Ⓒ Ⓓ	82 Ⓐ Ⓑ Ⓒ Ⓓ	122 Ⓐ Ⓑ Ⓒ Ⓓ
3 Ⓐ Ⓑ Ⓒ Ⓓ	43 Ⓐ Ⓑ Ⓒ Ⓓ	83 Ⓐ Ⓑ Ⓒ Ⓓ	123 Ⓐ Ⓑ Ⓒ Ⓓ
4 Ⓐ Ⓑ Ⓒ Ⓓ	44 Ⓐ Ⓑ Ⓒ Ⓓ	84 Ⓐ Ⓑ Ⓒ Ⓓ	124 Ⓐ Ⓑ Ⓒ Ⓓ
5 Ⓐ Ⓑ Ⓒ Ⓓ	45 Ⓐ Ⓑ Ⓒ Ⓓ	85 Ⓐ Ⓑ Ⓒ Ⓓ	125 Ⓐ Ⓑ Ⓒ Ⓓ
6 Ⓐ Ⓑ Ⓒ Ⓓ	46 Ⓐ Ⓑ Ⓒ Ⓓ	86 Ⓐ Ⓑ Ⓒ Ⓓ	126 Ⓐ Ⓑ Ⓒ Ⓓ
7 Ⓐ Ⓑ Ⓒ Ⓓ	47 Ⓐ Ⓑ Ⓒ Ⓓ	87 Ⓐ Ⓑ Ⓒ Ⓓ	127 Ⓐ Ⓑ Ⓒ Ⓓ
8 Ⓐ Ⓑ Ⓒ Ⓓ	48 Ⓐ Ⓑ Ⓒ Ⓓ	88 Ⓐ Ⓑ Ⓒ Ⓓ	128 Ⓐ Ⓑ Ⓒ Ⓓ
9 Ⓐ Ⓑ Ⓒ Ⓓ	49 Ⓐ Ⓑ Ⓒ Ⓓ	89 Ⓐ Ⓑ Ⓒ Ⓓ	129 Ⓐ Ⓑ Ⓒ Ⓓ
10 Ⓐ Ⓑ Ⓒ Ⓓ	50 Ⓐ Ⓑ Ⓒ Ⓓ	90 Ⓐ Ⓑ Ⓒ Ⓓ	130 Ⓐ Ⓑ Ⓒ Ⓓ
11 Ⓐ Ⓑ Ⓒ Ⓓ	51 Ⓐ Ⓑ Ⓒ Ⓓ	91 Ⓐ Ⓑ Ⓒ Ⓓ	131 Ⓐ Ⓑ Ⓒ Ⓓ
12 Ⓐ Ⓑ Ⓒ Ⓓ	52 Ⓐ Ⓑ Ⓒ Ⓓ	92 Ⓐ Ⓑ Ⓒ Ⓓ	132 Ⓐ Ⓑ Ⓒ Ⓓ
13 Ⓐ Ⓑ Ⓒ Ⓓ	53 Ⓐ Ⓑ Ⓒ Ⓓ	93 Ⓐ Ⓑ Ⓒ Ⓓ	133 Ⓐ Ⓑ Ⓒ Ⓓ
14 Ⓐ Ⓑ Ⓒ Ⓓ	54 Ⓐ Ⓑ Ⓒ Ⓓ	94 Ⓐ Ⓑ Ⓒ Ⓓ	134 Ⓐ Ⓑ Ⓒ Ⓓ
15 Ⓐ Ⓑ Ⓒ Ⓓ	55 Ⓐ Ⓑ Ⓒ Ⓓ	95 Ⓐ Ⓑ Ⓒ Ⓓ	135 Ⓐ Ⓑ Ⓒ Ⓓ
16 Ⓐ Ⓑ Ⓒ Ⓓ	56 Ⓐ Ⓑ Ⓒ Ⓓ	96 Ⓐ Ⓑ Ⓒ Ⓓ	136 Ⓐ Ⓑ Ⓒ Ⓓ
17 Ⓐ Ⓑ Ⓒ Ⓓ	57 Ⓐ Ⓑ Ⓒ Ⓓ	97 Ⓐ Ⓑ Ⓒ Ⓓ	137 Ⓐ Ⓑ Ⓒ Ⓓ
18 Ⓐ Ⓑ Ⓒ Ⓓ	58 Ⓐ Ⓑ Ⓒ Ⓓ	98 Ⓐ Ⓑ Ⓒ Ⓓ	138 Ⓐ Ⓑ Ⓒ Ⓓ
19 Ⓐ Ⓑ Ⓒ Ⓓ	59 Ⓐ Ⓑ Ⓒ Ⓓ	99 Ⓐ Ⓑ Ⓒ Ⓓ	139 Ⓐ Ⓑ Ⓒ Ⓓ
20 Ⓐ Ⓑ Ⓒ Ⓓ	60 Ⓐ Ⓑ Ⓒ Ⓓ	100 Ⓐ Ⓑ Ⓒ Ⓓ	140 Ⓐ Ⓑ Ⓒ Ⓓ
21 Ⓐ Ⓑ Ⓒ Ⓓ	61 Ⓐ Ⓑ Ⓒ Ⓓ	101 Ⓐ Ⓑ Ⓒ Ⓓ	141 Ⓐ Ⓑ Ⓒ Ⓓ
22 Ⓐ Ⓑ Ⓒ Ⓓ	62 Ⓐ Ⓑ Ⓒ Ⓓ	102 Ⓐ Ⓑ Ⓒ Ⓓ	142 Ⓐ Ⓑ Ⓒ Ⓓ
23 Ⓐ Ⓑ Ⓒ Ⓓ	63 Ⓐ Ⓑ Ⓒ Ⓓ	103 Ⓐ Ⓑ Ⓒ Ⓓ	143 Ⓐ Ⓑ Ⓒ Ⓓ
24 Ⓐ Ⓑ Ⓒ Ⓓ	64 Ⓐ Ⓑ Ⓒ Ⓓ	104 Ⓐ Ⓑ Ⓒ Ⓓ	144 Ⓐ Ⓑ Ⓒ Ⓓ
25 Ⓐ Ⓑ Ⓒ Ⓓ	65 Ⓐ Ⓑ Ⓒ Ⓓ	105 Ⓐ Ⓑ Ⓒ Ⓓ	145 Ⓐ Ⓑ Ⓒ Ⓓ
26 Ⓐ Ⓑ Ⓒ Ⓓ	66 Ⓐ Ⓑ Ⓒ Ⓓ	106 Ⓐ Ⓑ Ⓒ Ⓓ	146 Ⓐ Ⓑ Ⓒ Ⓓ
27 Ⓐ Ⓑ Ⓒ Ⓓ	67 Ⓐ Ⓑ Ⓒ Ⓓ	107 Ⓐ Ⓑ Ⓒ Ⓓ	147 Ⓐ Ⓑ Ⓒ Ⓓ
28 Ⓐ Ⓑ Ⓒ Ⓓ	68 Ⓐ Ⓑ Ⓒ Ⓓ	108 Ⓐ Ⓑ Ⓒ Ⓓ	148 Ⓐ Ⓑ Ⓒ Ⓓ
29 Ⓐ Ⓑ Ⓒ Ⓓ	69 Ⓐ Ⓑ Ⓒ Ⓓ	109 Ⓐ Ⓑ Ⓒ Ⓓ	149 Ⓐ Ⓑ Ⓒ Ⓓ
30 Ⓐ Ⓑ Ⓒ Ⓓ	70 Ⓐ Ⓑ Ⓒ Ⓓ	110 Ⓐ Ⓑ Ⓒ Ⓓ	150 Ⓐ Ⓑ Ⓒ Ⓓ
31 Ⓐ Ⓑ Ⓒ Ⓓ	71 Ⓐ Ⓑ Ⓒ Ⓓ	111 Ⓐ Ⓑ Ⓒ Ⓓ	151 Ⓐ Ⓑ Ⓒ Ⓓ
32 Ⓐ Ⓑ Ⓒ Ⓓ	72 Ⓐ Ⓑ Ⓒ Ⓓ	112 Ⓐ Ⓑ Ⓒ Ⓓ	152 Ⓐ Ⓑ Ⓒ Ⓓ
33 Ⓐ Ⓑ Ⓒ Ⓓ	73 Ⓐ Ⓑ Ⓒ Ⓓ	113 Ⓐ Ⓑ Ⓒ Ⓓ	153 Ⓐ Ⓑ Ⓒ Ⓓ
34 Ⓐ Ⓑ Ⓒ Ⓓ	74 Ⓐ Ⓑ Ⓒ Ⓓ	114 Ⓐ Ⓑ Ⓒ Ⓓ	154 Ⓐ Ⓑ Ⓒ Ⓓ
35 Ⓐ Ⓑ Ⓒ Ⓓ	75 Ⓐ Ⓑ Ⓒ Ⓓ	115 Ⓐ Ⓑ Ⓒ Ⓓ	155 Ⓐ Ⓑ Ⓒ Ⓓ
36 Ⓐ Ⓑ Ⓒ Ⓓ	76 Ⓐ Ⓑ Ⓒ Ⓓ	116 Ⓐ Ⓑ Ⓒ Ⓓ	156 Ⓐ Ⓑ Ⓒ Ⓓ
37 Ⓐ Ⓑ Ⓒ Ⓓ	77 Ⓐ Ⓑ Ⓒ Ⓓ	117 Ⓐ Ⓑ Ⓒ Ⓓ	157 Ⓐ Ⓑ Ⓒ Ⓓ
38 Ⓐ Ⓑ Ⓒ Ⓓ	78 Ⓐ Ⓑ Ⓒ Ⓓ	118 Ⓐ Ⓑ Ⓒ Ⓓ	158 Ⓐ Ⓑ Ⓒ Ⓓ
39 Ⓐ Ⓑ Ⓒ Ⓓ	79 Ⓐ Ⓑ Ⓒ Ⓓ	119 Ⓐ Ⓑ Ⓒ Ⓓ	159 Ⓐ Ⓑ Ⓒ Ⓓ
40 Ⓐ Ⓑ Ⓒ Ⓓ	80 Ⓐ Ⓑ Ⓒ Ⓓ	120 Ⓐ Ⓑ Ⓒ Ⓓ	160 Ⓐ Ⓑ Ⓒ Ⓓ

NOTATIONS

(a, b)	$\{ x : a < x < b \}$
$[a, b)$	$\{ x : a \leq x < b \}$
$(a, b]$	$\{ x : a < x \leq b \}$
$[a, b]$	$\{ x : a \leq x \leq b \}$
$\gcd(m, n)$	greatest common divisor of two integers m and n
$\text{lcm}(m, n)$	least common multiple of two integers m and n
$[x]$	greatest integer m such that $m \leq x$
$m \equiv k \pmod{n}$	m and k are congruent modulo n (m and k have the same remainder when divided by n, or equivalently, $m - k$ is a multiple of n)
f^{-1}	inverse of an invertible function f; (not to be read as $\frac{1}{f}$)
$\lim_{x \to a^+} f(x)$	right-hand limit of $f(x)$; limit (if it exists) of $f(x)$ as x approaches a from the right
$\lim_{x \to a^-} f(x)$	left-hand limit of $f(x)$; limit (if it exists) of $f(x)$ as x approaches a from the left
$p \to q$	statement p implies statement q; if p, then q
$p \leftrightarrow q$	statements p and q are logically equivalent; p if and only if q
$\sim p$	negation of the statement p; not p
$p \wedge q$	conjunction of statements p and q; p and q
$p \vee q$	disjunction of statements p and q; p or q
$p \veebar q$	exclusive or for statements p and q; p or q but not both
$B \leftarrow C$	(in a computer algorithm) value of the variable B is replaced by the current value of the variable C
$\varnothing$	the empty set
$x \in S$	x is an element of set S
$S \subset T$	set S is a proper subset of set T
$S \subseteq T$	either set S is a proper subset of set T or $S = T$
$\overline{S}$	complement of set S; the set of all elements not in S that are in some specified universal set
$T \setminus S$	relative complement of set S in set T, i.e., the set of all elements of T that are not elements of S
$S \cup T$	union of sets S and T
$S \cap T$	intersection of sets S and T

FORMULAS

Sum
$$\sin(x + y) = \sin x \cos y + \cos x \sin y$$
$$\cos(x + y) = \cos x \cos y - \sin x \sin y$$

Half-Angle   (sign depends on the quadrant of $\frac{\theta}{2}$)

$$\sin\frac{\theta}{2} = \pm\sqrt{\frac{1 - \cos\theta}{2}}; \qquad \cos\frac{\theta}{2} = \pm\sqrt{\frac{1 + \cos\theta}{2}}$$

Range of inverse trigonometric functions

$$\sin^{-1}x \ [-\pi/2, \pi/2]; \qquad \cos^{-1}x \ [0, \pi]; \qquad \tan^{-1}x \ (-\pi/2, \pi/2)$$

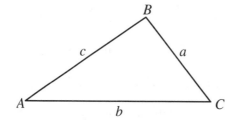

Law of Sines $\dfrac{\sin A}{\sin B} = \dfrac{a}{b}$

Law of Cosines $c^2 = a^2 + b^2 - 2ab(\cos C)$

DeMoivre's Theorem $(\cos\theta + i\sin\theta)^k = \cos(k\theta) + i\sin(k\theta)$

Coordinate Transformation

Rectangular (x, y) to polar (r, θ) $r^2 = x^2 + y^2$; $\tan\theta = \dfrac{y}{x}$ if $x \neq 0$

Polar (r, θ) to rectangular (x, y) $x = r\cos\theta$; $y = r\sin\theta$

Distance from point (x_1, y_1) to line $Ax + By + C = 0$ $d = \dfrac{|Ax_1 + By_1 + C|}{\sqrt{A^2 + B^2}}$

Volume

Sphere: radius r $\qquad$ $V = \frac{4}{3}\pi r^3$

Right circular cone: height h, base of radius r $\qquad$ $V = \frac{1}{3}\pi r^2 h$

Right circular cylinder: height h, base of radius r $\qquad$ $V = \pi r^2 h$

Pyramid: height h, base of area B $\qquad$ $V = \frac{1}{3}Bh$

Right prism: height h, base of area B $\qquad$ $V = Bh$

Surface Area

Sphere: radius r $\qquad$ $A = 4\pi r^2$

Lateral surface area of right circular cone: radius r, slant height s $\qquad$ $A = \pi rs$

Differentiation

$$(f(x)g(x))' = f'(x)g(x) + f(x)g'(x) \qquad (f[\,g(x)])' = f'[\,g(x)]\,g'(x)$$

$$\left(\frac{f(x)}{g(x)}\right)' = \frac{f'(x)g(x) - f(x)g'(x)}{(g(x))^2} \qquad \text{if } g(x) \neq 0$$

Integration by Parts $\displaystyle \int u\,dv = uv - \int v\,du$

DEFINITIONS

Linear Algebra

A vector $\mathbf{u}$ is a <u>linear combination</u> of the vectors $\mathbf{v}_1, \mathbf{v}_2, \mathbf{v}_3, \ldots, \mathbf{v}_n$ if there exist real numbers $a_1, a_2, a_3, \ldots, a_n$ such that $\mathbf{u} = a_1\mathbf{v}_1 + a_2\mathbf{v}_2 + a_3\mathbf{v}_3 + \ldots + a_n\mathbf{v}_n$.

The <u>linear span</u> of the vectors $\mathbf{v}_1, \mathbf{v}_2, \mathbf{v}_3, \ldots, \mathbf{v}_n$ is the set of all linear combinations of $\mathbf{v}_1, \mathbf{v}_2, \mathbf{v}_3, \ldots, \mathbf{v}_n$.

The vectors $\mathbf{v}_1, \mathbf{v}_2, \mathbf{v}_3, \ldots, \mathbf{v}_n$ are <u>linearly independent</u> if $a_1\mathbf{v}_1 + a_2\mathbf{v}_2 + a_3\mathbf{v}_3 + \ldots + a_n\mathbf{v}_n = \mathbf{0}$ (the zero vector) implies that $a_1 = a_2 = a_3 = \ldots = a_n = 0$.

The set of vectors $\mathbf{v}_1, \mathbf{v}_2, \mathbf{v}_3, \ldots, \mathbf{v}_k$ forms a <u>basis</u> for a subspace W of $\mathbf{R}^n$ if $\mathbf{v}_1, \mathbf{v}_2, \mathbf{v}_3, \ldots, \mathbf{v}_k$ are linearly independent and their linear span is equal to W.

The <u>dimension</u> of a subspace W of $\mathbf{R}^n$ is the number of vectors in a basis for W.

Discrete Mathematics

A relation $\Re$ on a set S is

<u>reflexive</u> if $x \Re x$ for all $x \in S$

<u>symmetric</u> if $x \Re y \Rightarrow y \Re x$ for all $x, y \in S$

<u>transitive</u> if $(x \Re y$ and $y \Re z) \Rightarrow x \Re z$ for all $x, y, z \in S$

<u>antisymmetric</u> if $(x \Re y$ and $y \Re z) \Rightarrow x = y$ for all $x, y, \in S$

An <u>equivalence</u> relation is a reflexive, symmetric, and transitive relation.

MATHEMATICS: CONTENT KNOWLEDGE

Directions: Each of the questions or incomplete statements below is followed by four suggested answers or completions. Select the one that is best in each case and then fill in the corresponding lettered space on the answer sheet with a heavy, dark mark so that you cannot see the letter.

1. Into at most how many disjoint regions can three lines divide the plane?

   (A) 6
   (B) 7
   (C) 8
   (D) 9

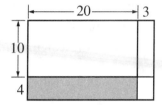

2. The area of a rectangular region with adjacent sides of lengths a and b, respectively, can be used to represent the product ab. What product is represented by the shaded rectangular region in the figure above?

   (A) 4×3
   (B) 4×20
   (C) 10×4
   (D) 10×20

3. José is considering what sandwich to buy for lunch. He has a choice of 2 different types of bread, 3 different types of cheese, and 4 different types of meat. For any sandwich, he must choose one type of bread, and either one type of cheese *or* one type of meat, or one type of cheese *and* one type of meat. How many different sandwiches can he choose?

   (A) 14
   (B) 24
   (C) 38
   (D) 48

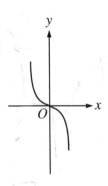

NUMBER OF DAYS TAKEN TO COMPLETE
A CERTAIN ASSIGNMENT

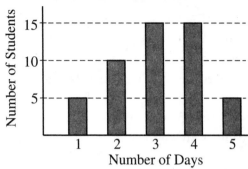

4. If the graph of the function $y = f(x)$ is shown above, which of the following could be the graph of $y = f(x+1) - 2$?

(A)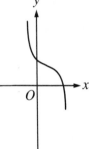

(B)

(C)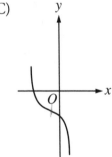

(D)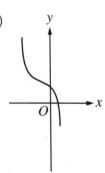

5. An assignment was given to 50 middle school students to complete. The graph above shows a frequency distribution of the number of days it took the students to complete the assignment. What is the mean number of days it took for a student to complete the assignment?

(A) 2.8
(B) 3.0
(C) 3.1
(D) 3.3

6. If $a < b$, which of the following <u>must</u> be true?

(A) $|a-b| > 0$
(B) $|a| < |b|$
(C) $|a| \neq |b|$
(D) $|b| - |a| < |b-a|$

7. Which of the following sets is finite?

(A) All integer multiples of $\frac{1}{2}$ that are greater than 0 and less than 1,000
(B) All rational numbers that are greater than 7 and less than 8
(C) All rational numbers with decimal expansions that terminate
(D) All irrational numbers

8. The graph of the axis of symmetry of the parabola $y = 2x^2 - 11x + 3$ is best described by which of the following equations?

   (A) $x = 2$
   (B) $x = 2.25$
   (C) $x = 2.5$
   (D) $x = 2.75$

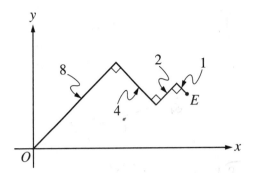

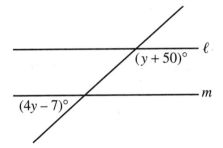

9. In the figure above, lines ℓ and m are parallel. What is the value of y ?

   (A) 19

   (B) $\dfrac{137}{5}$

   (C) $\dfrac{223}{5}$

   (D) $\dfrac{137}{3}$

10. In the figure above, what is the distance from point E to the origin?

    (A) 10
    (B) 15
    (C) $5\sqrt{5}$
    (D) $\dfrac{5\sqrt{10}}{2}$

11. If $\tan x = \dfrac{2}{3}$ and $\pi < x < \dfrac{3\pi}{2}$, then $\cos x =$

    (A) $\dfrac{3}{\sqrt{13}}$

    (B) $\dfrac{2}{\sqrt{13}}$

    (C) $-\dfrac{2}{\sqrt{13}}$

    (D) $-\dfrac{3}{\sqrt{13}}$

12. A continuous random variable X has mean 60.0 and standard deviation 8. What value does the random variable have 2.5 standard deviations above the mean?

    (A) 62.5
    (B) 70.5
    (C) 76.0
    (D) 80.0

13. If A is a 2×2 matrix and $A\begin{pmatrix} x \\ y \end{pmatrix} = 3\begin{pmatrix} x \\ y \end{pmatrix}$ for all real x and y, then $A =$

(A) $\begin{pmatrix} 3 & 0 \\ 0 & 3 \end{pmatrix}$

(B) $\begin{pmatrix} 3 & 0 \\ 0 & 0 \end{pmatrix}$

(C) $\begin{pmatrix} 3 & 3 \\ 3 & 3 \end{pmatrix}$

(D) $\begin{pmatrix} 0 & 3 \\ 3 & 0 \end{pmatrix}$

14. What is printed in STEP 4 when the program below is executed?

STEP 1: $X = 100, Y = 10$

STEP 2: If $((X \geq 100)$ or $(X < 20))$

Then $X \leftarrow (X + 1)$

Else $X \leftarrow (X - 1)$

STEP 3: If $((X > Y)$ and $(Y < 10))$

Then $X \leftarrow (X + Y)$

Else $X \leftarrow (X - Y)$

STEP 4: Print the value of X

(A) 89
(B) 91
(C) 109
(D) 111

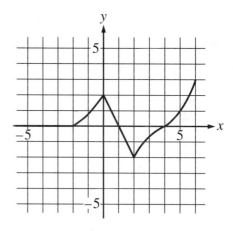

15. The graph of the integrable function $y = f(x)$ is shown above. Which of the following is the best approximation of $\int_{-2}^{4} f(x)\,dx$?

(A) 7
(B) 5
(C) 2
(D) 0

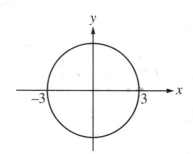

16. The circle in the figure above is centered at the origin and has radius 3. How many points with integral coordinates lie inside or on the boundary of the circular region?

(A) 26
(B) 27
(C) 29
(D) 32

17. In a league that has 5 teams, each team plays each of the other teams twice in a season. What is the total number of games played during the season?

    (A) 10
    (B) 12
    (C) 20
    (D) 24

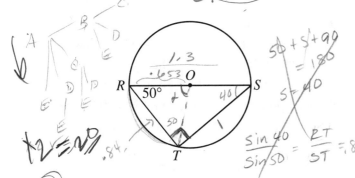

18. In the circle above with center O, what is the degree measure of arc RT ?

    (A) 40°
    (B) 50°
    (C) 80°
    (D) 90°

19. Beginning at time $t = 0$, the number of rabbits in a certain population at time t years is modeled by the function $f(t) = \dfrac{10,000}{10 + 50e^{-0.5t}}$.

    According to this model, which of the following best describes how the size of the population changes over time?

    (A) It increases without bound.
    (B) It increases for several years, then levels off.
    (C) It increases for several years, then decreases to zero.
    (D) It increases then decreases in cycles.

20. In a history class, a teacher has given the students the following schedule for determining their grades for the course:

3 tests (20% each)	60%
1 paper	20%
2 quizzes (5% each)	10%
Class participation	10%

    Jane has the following percentage grades:

Test scores	93, 82, 89
Paper	95
Quizzes	86, 96

    What will her class participation grade need to be in order to guarantee a grade of 90% for the course?

    (A) 90
    (B) 91
    (C) 92
    (D) 93

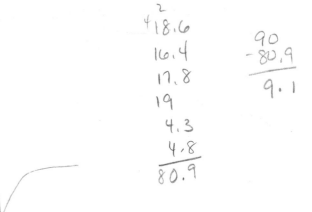

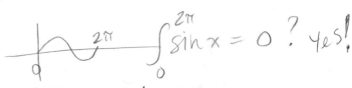

$$x^3 - 3x^2 + 2x - 6$$

21. What is the value of $\lim\limits_{x \to 3} \dfrac{x^3 - 3x^2 + 2x - 6}{x - 3}$?

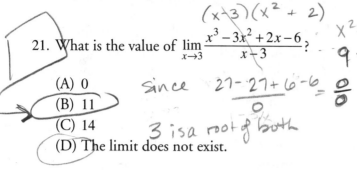

(x-3)(x² + 2)   x² + 2
9 + 2 = 11

(A) 0
(B) 11
(C) 14
(D) The limit does not exist.

since 27 - 27 + 6 - 6 = 0/0
                0

3 is a root of both

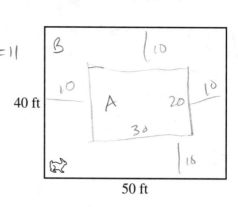

40 ft

50 ft

22. The Richter scale is a base 10 logarithmic scale used to measure the magnitude of earthquakes; i.e., an earthquake measuring 7 is ten times as strong as an earthquake measuring 6. An earthquake that measures 6.8 on the Richter scales has a magnitude that is approximately what percent of an earthquake measuring 6.6 ?

(A) 103%
(B) 120%
(C) 158%
(D) 200%

7R = 10 × 6R
6.8R = ___ ×6.6R
log₁₀ B = X
            10ˣ = B

24. A rabbit is hopping around in a fenced-off flat rectangular field with dimensions as shown above. If the position of the rabbit is uniformly random throughout the field, what is the probability that the rabbit is 10 feet or more away from the fence at any given time?

(A) 0.80
(B) 0.75
(C) 0.60
(D) 0.30

P(A) = 30×20 = 600

P(B) = (40×50) - P(A)
     = 2,000 - 600
     = 1400

P(A)/P(TOTAL) = 600/2000 = 30%

23. The surface area of a sphere is approximately 1,500 square inches. What is the approximate volume of the sphere, in cubic inches?

(A) 500
(B) 1,500
(C) 3,500
(D) 5,500

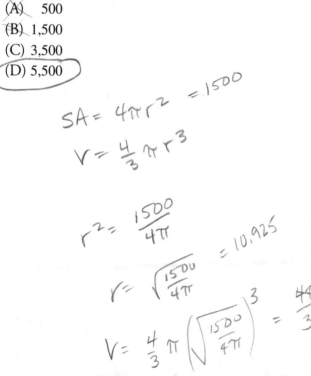

SA = 4πr² = 1500

V = 4/3 πr³

r² = 1500/4π

r = √(1500/4π) = 10.925

V = 4/3 π (√(1500/4π))³ = 4π/3 · 1500^(3/2)/4π^(1/2)

Log₁₀ B₁ = 7
Log₁₀ B₂ = 6

B₁ = 10⁷/10⁶ = 10
B₂   10

Log₁₀ B₃ = 6.8
Log₁₀ B₄ = 6.6

10^6.8 / 10^6.6

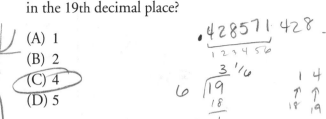

25. If a differentiable function is strictly increasing on the real numbers, which of the following statements about the function's derivative must be true?

(A) It is always positive and strictly increasing.

(B) It is always positive but need not be strictly increasing.

(C) It need not always be positive but is always strictly increasing.

(D) It need not always be positive and need not always be strictly increasing.

26. Which of the following sets is NOT closed under division?

(A) Integers, excluding 0

(B) Rational numbers, excluding 0

(C) Real numbers, excluding 0

(D) Complex numbers, excluding 0

27. Given $f(x) = 4x$, $g(x) = x^2 + 1$, and $h(x) = \dfrac{1}{x}$, for what value or values of x is $h(f(g(x))) = \dfrac{1}{4}$?

(A) 0

(B) $\dfrac{\sqrt{3}}{2}$

(C) $\pm\dfrac{\sqrt{3}}{2}$

(D) $\pm i\dfrac{\sqrt{15}}{15}$

28. If $\dfrac{3}{7}$ is expressed in decimal form, what digit is in the 19th decimal place?

(A) 1

(B) 2

(C) 4

(D) 5

29. The following is a computer algorithm using the numerical variables N and X. The function INT (x) is defined to be the greatest integer less than or equal to x.

STEP 1: Read a numerical value of N

STEP 2: $X \leftarrow \dfrac{\text{INT}(100N + 0.5)}{100}$

STEP 3: Print X

STEP 4: End

If the value of N read in STEP 1 is 6,735.8291, what is printed in STEP 3 when the algorithm is executed?

(A) 6735

(B) 6735.83

(C) 6736

(D) 6736.32

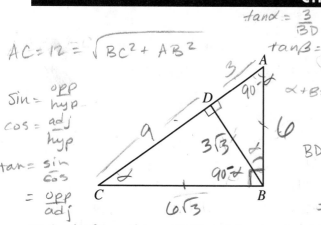

Handwritten notes (top):

$BD = \dfrac{3}{\tan\alpha} = \dfrac{9}{\tan\beta}$
$\dfrac{3}{9} = \dfrac{\tan\alpha}{\tan\beta} = \dfrac{1}{3}$
$\tan\beta = 3\tan\alpha$
$\beta = \tan^{-1}(3\tan\alpha)$
$\alpha + \beta = 90$
$\alpha + \tan^{-1}(3\tan\alpha) = 90$

$\tan\alpha = \dfrac{3}{BD}$

$\tan\beta = \dfrac{9}{BD}$

$\alpha + \beta = 90$

$AC = 12 = \sqrt{BC^2 + AB^2}$

$\sin = \dfrac{opp}{hyp}$

$\cos = \dfrac{adj}{hyp}$

$\tan = \dfrac{\sin}{\cos} = \dfrac{opp}{adj}$

$BD = \sqrt{36-9} = \sqrt{27} = 3\sqrt{3}$

32. If the real-valued function $f(x) = x^n$, where n is a positive integer, has neither a local maximum nor a local minimum, which of the following <u>must</u> be true?

(A) n is even.
(B) n is odd.
(C) n is a multiple of 3.
(D) n can be any positive integer.

$n=1$
$n-1=0$
$f'(x) = nx^{n-1}$
cannot $= 0$

30. In the figure above, ABC is a right triangle with right angle at B. $\overline{BD}$ is perpendicular to $\overline{AC}$. If $CD = 9$ and $AD = 3$, then $BC + AB + BD =$

$6 + 9\sqrt{3}$

(A) $6 + 9\sqrt{3}$
(B) $6\sqrt{3} + 9$
(C) $2\sqrt{2} + 7$
(D) $4 + 3\sqrt{2}$

$\sin\alpha = \dfrac{AB}{12} = \dfrac{3}{AB} \Rightarrow AB^2 = 36 \quad AB = 6$

$\dfrac{\sin\alpha}{\sin(90-\alpha)} = \dfrac{AB}{BC}$

$\sin(90-\alpha) = \dfrac{9}{BC}$

$BC = \sqrt{144-36} = \sqrt{108} = 3\sqrt{12} = 6\sqrt{3}$

31. In the course of solving an equation, which of the following procedures can result in an equation that yields a real root that does not satisfy the original equation?

(A) Subtracting the same number from both sides of the equation
(B) Raising both sides of the equation to the third power
(C) Squaring both sides of the equation
(D) Dividing both sides of the equation by a nonzero number

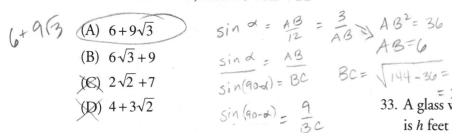

$A_1 = \frac{1}{3}$
$A_2 = A_{11}$

33. A glass window is composed of a rectangle that is h feet long by d feet wide and a semicircle with diameter d, as shown in the figure above.

If the area of the semicircular region is $\dfrac{1}{3}$ of the total area of the window, what is the ratio of h to d?

(A) π to 2
(B) π to 3
(C) π to 4
(D) It cannot be determined from the information given.

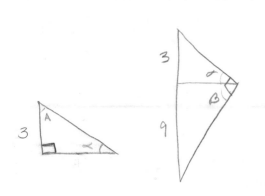

$\dfrac{\sin\alpha}{\sin\beta} = \dfrac{1}{3}$

$\sin\alpha = \dfrac{3}{AB}$

$\sin\beta = \dfrac{9}{CB} = 3\sin\alpha$

$\sin\beta = 3\sin\alpha$

$\sin(90-\alpha) = 3\sin\alpha$

$\sin 90 - \sin\alpha = 3\sin\alpha$

$AB = \dfrac{3}{\sin\alpha}$

$CB = \dfrac{9}{3\sin\alpha} = \dfrac{1}{3\sin\alpha}$

$1 = 4\sin\alpha$
$\sin\alpha = .25$
$\alpha = 14.48°$
$\beta = 48.59$
63

$AB^2 + CB^2 = 12^2$

$\dfrac{9}{\sin^2\alpha} + \dfrac{1}{9\sin^2\alpha} = 144 = \dfrac{81}{9\sin^2\alpha} + \dfrac{1}{9\sin^2\alpha} = 144 \quad \dfrac{82}{9\sin^2\alpha} = 144$

34. Which of the following is the solution set of the matrix equation $\begin{bmatrix} 1 & -1 \\ 1 & 3 \end{bmatrix}\begin{bmatrix} x \\ y \end{bmatrix} = 2\begin{bmatrix} x \\ y \end{bmatrix}$?

(A) $\left\{ \begin{bmatrix} 0 \\ 0 \end{bmatrix} \right\}$

(B) $\left\{ \begin{bmatrix} 0 \\ 0 \end{bmatrix}, \begin{bmatrix} 1 \\ -1 \end{bmatrix} \right\}$

(C) $\left\{ \begin{bmatrix} -c \\ c \end{bmatrix},$ where c is any real number $\right\}$

(D) There are no solutions to the matrix equation.

handwritten: $3 - 3 = 6$ $x - y = 2x$ $-x - y = 0$ $x + 3y = 2y$ $+ x + y = 0$ $x - y$ $-x = y$ $-y = x$ ✓

35. Karen opened a bank account for her son on his 1st birthday with a $100 deposit. After that, $50 was deposited in the account on each birthday. No withdrawals and no other deposits were made until his 11th birthday. The bank pays 8% interest per year, compounded annually. Which of the following recursive sequences models the amount of money in the account after n years, $1 \le n \le 10$?

(A) $A(0) = 100$
$A(n) = 0.08\,A(n-1) + 50$

(B) $A(0) = 100$
$A(n) = 1.08\,A(n-1) + 50$

(C) $A(0) = 100$
$A(n) = 0.08\,[A(n-1) + 50]$

(D) $A(0) = 0$
$A(n) = 1.08\,[A(n-1) + 100] + 50$

handwritten: 2ND B-day $A_1 = 1.08(100) + 50$ $1.08(100 + 100) + 50$ $100 + 1.08$

36. If $f(x) = x^5 - 7x^3 + 6x^2 - 2x + 17$, for how many values of x is the derivative of $f(x)$ equal to zero?

(A) Two
(B) Three
(C) Four
(D) Five

handwritten: $f'(x) = 5x^4 - 21x^2 + 12x - 2$ $x = 1.713$ $= -2.29$

handwritten table:

$\times$	a	b
a	b	a
b	a	a

handwritten: $a \times a = b$ $a \times b = a$ $b \times a = a$ $b \times b = a$
$a \times b = b \times a$ $a \times (b \times a) = (a \times b) \times a$

37. The table above defines an operation $\times$ on the set $S = \{a, b\}$. Which of the following statements about $\times$ is true?

(A) It is neither associative nor commutative.
(B) It is associative but not commutative.
(C) It is commutative but not associative.
(D) It is both associative and commutative.

38. Which of the following is an equation of an asymptote to the hyperbola $(x-1)^2 - (y+2)^2 = 1$?

(A) $y = x - 3$
(B) $y = -x - 3$
(C) $y = x - 1$
(D) $y = x + 1$

handwritten: $\dfrac{(x-h)^2}{a^2} - \dfrac{(y-k)^2}{b^2} = 1$ center @ (h, k) $(h, k) = (1, -2)$ $a = b = 1$

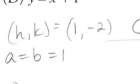

handwritten: $y = x - 3$ $y = -x - 1$ comm. $a \times b = b \times a$ ass. $a \times (b \times c) = (a \times b) \times c$

handwritten bottom: $144 \times 9 \sin 2\alpha = 82$ $\sin 2\alpha = .0633$ $\sin \alpha = .251$ $\alpha = 14.7°$ $\beta = 49.6$ NO!

39. In the *xy*-plane, what is the radius of the circle described by the equation $x^2 + 2x + y^2 = 0$?

(A) 1
(B) 2
(C) 3
(D) 4

40. What is the period of the function

$$y = \frac{1}{3}\sin\left(\frac{1}{2}x + \frac{\pi}{3}\right)?$$

(A) $\frac{1}{3}$

(B) $\frac{\pi}{6}$

(C) π

(D) 4π

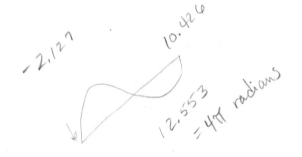

SCORES ON AN ALGEBRA I TEST

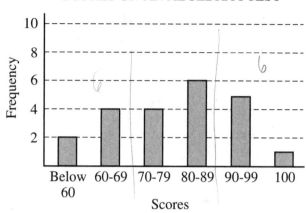

41. Which of the following statistics about the algebra I test scores can be determined using only the information above?

I. The range of scores
II. The median score
III. The average (arithmetic mean) score

(A) None
(B) I only
(C) II only
(D) III only

42. The function $f(x) = x^3 + 1$ is one-to-one; therefore, the inverse function $y = f^{-1}(x)$ exists. Which of the following is an equation for $y = f^{-1}(x)$?

    (A) $y = \dfrac{1}{x^3 + 1}$

    (B) $y = \sqrt[3]{x - 1}$

    (C) $y = \sqrt[3]{x} - 1$

    (D) $y = \dfrac{1}{\sqrt[3]{x - 1}}$

[Handwritten: $f(x) = x^3 + 1$ $\quad y = x^3 + 1$ $\quad f^{-1}(x) = ?$ $\quad$ inverse function reverses role of x & y $\quad x = y^3 + 1$ (solve for y) $\quad y^3 = x - 1$ $\quad y = \sqrt[3]{x-1} = f^{-1}(x)$]

43. Consider the set S of 2×2 matrices, all of whose entries are nonzero real numbers. Which of the following properties is satisfied by S under multiplication?

    I. S is closed.

    II. S is commutative.

    III. S contains an identity.

    (A) None

    (B) I only

    (C) I and II only

    (D) I and III only

[Handwritten work: $\begin{bmatrix}1&1\\1&1\end{bmatrix}\begin{bmatrix}a\\-a\end{bmatrix} = \begin{bmatrix}0\\0\end{bmatrix}$; $\begin{bmatrix}1&2\\2&1\end{bmatrix}\begin{bmatrix}3&1\\0&3\end{bmatrix} = \begin{bmatrix}3&7\\7&5\end{bmatrix} = \begin{bmatrix}5\end{bmatrix}$ NOT =; $\begin{bmatrix}a&b\\c&d\end{bmatrix}\begin{bmatrix}e&f\\g&h\end{bmatrix} = \begin{bmatrix}i&j\\k&e\end{bmatrix}$ NO]

44. A sequence is defined recursively by
$$a_n = \begin{cases} 1 & \text{if } n = 1 \\ a_{n-1} + n & \text{if } n > 1 \end{cases}.$$

Which of the following is a closed-form representation of the sequence?

    (A) $a_n = \dfrac{(n-1)n}{2}$

    (B) $a_n = \dfrac{n+1}{2}$

    (C) $a_n = \dfrac{n(n+1)}{2}$

    (D) $a_n = n^2 - (n-1)(n+1)$

[Handwritten: $a_1 = 1$, $a_2 = 3$, $a_3 = 6$; 3×4]

45. For which of the following values of x is
$$f(x) = \frac{\sqrt{x^2 + 4}}{x^3 + x^2 - 5x + 3}$$
undefined?

    I. -3

    II. -2

    III. 1

    (A) I only

    (B) II only

    (C) I and III only

    (D) I, II, and III

[Handwritten: $-27 + 9 + 15 + 3$; $-8 + 4 + 10 + 3 =$; $1 + 1 - 5 + 3$; $9 - (3-1)(3+1) \quad 9 - (2)(4) = 3$]

[Handwritten work at bottom: for any B, $[A_s][B_s] = [B_s]$; $\begin{bmatrix}a_1&a_2\\a_3&a_4\end{bmatrix}\begin{bmatrix}1&1\\1&1\end{bmatrix} = \begin{bmatrix}1&1\\1&1\end{bmatrix}$; $\begin{bmatrix}a_1+a_2&a_1+a_2\\a_3+a_4&a_3+a_4\end{bmatrix} = \begin{bmatrix}1&1\\1&1\end{bmatrix}$; $\begin{bmatrix}a_1&a_2\\a_3&a_4\end{bmatrix}\begin{bmatrix}1&2\\3&4\end{bmatrix} = \begin{bmatrix}a_1+3a_2&2a_1+4a_2\\ & \end{bmatrix} = \begin{bmatrix}1&2\\3&4\end{bmatrix}$]

[Handwritten: so $a_1 + a_2 = 1$ ∴ $a_2 = \emptyset$; $a_1 + 3a_2 = 1$; ∴ $[A]$ is not in S]

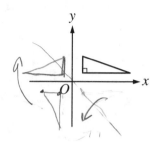

46. The triangle shown in the plane above is to be transformed in the plane by being reflected about the line $y = -x$ and then rotated 90° clockwise about the origin. Which of the following could be the result of this transformation?

(A)

(B)

(C)

(D)

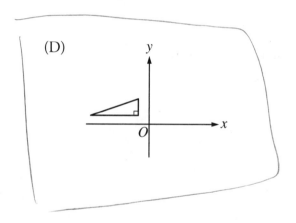

47. Given $\sum_{i=1}^{n} i^2 = \frac{n(n+1)(2n+1)}{6}$, what is the

value of $\sum_{i=50}^{75} i^2$?

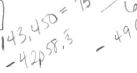

(A) 97,924

(B) 100,525

(C) 103,025

(D) 143,450

48. At how many points in the xy-plane do the graphs of

$y = 0.25x^4 + 0.4x^3 - 1.2x^2 - 0.75x - 0.25$

and $y = 0.5x - 2$ intersect?

(A) None

(B) Two

(C) Three

(D) Four

49. A standard six-sided die is weighted so that the probabilities of throwing 2, 3, 4, 5, or 6 are equal and the probability of throwing a 1 is twice the probability of throwing a 2. If the die is thrown twice, what is the probability that the sum of the numbers thrown will be 4?

(A) $\frac{1}{12}$

(B) $\frac{5}{49}$

(C) $\frac{3}{7}$

(D) $\frac{5}{11}$

50. The function $y = f(x) \cdot g(x)$, where

$$f(x) = x \text{ and } g(x) = \begin{cases} -1 & \text{if } x < 0 \\ 0 & \text{if } x = 0 \\ 1 & \text{if } x > 0 \end{cases},$$

is equivalent to which of the following functions?

(A) $y = f(x)$

(B) $y = f(-x)$

(C) $y = -|f(x)|$

(D) $y = f(|x|)$

Chapter 7

Practice Test, *Mathematics: Proofs, Models, and Problems, Part 1*

▶   ▶   ▶   ▶   ▶   ▶   ▶   ▶   ▶   ▶   ▶   ▶

Now that you have studied the content topics and have worked through strategies relating to the *Proofs, Models, and Problems* test, you should take the following practice test. You will probably find it helpful to simulate actual testing conditions, giving yourself 60 minutes to work on the questions. You can use the response pages provided if you wish.

Keep in mind that the test you take at an actual administration will have different questions. You should not expect your level of performance to be exactly the same as when you take the test at an actual administration, since numerous factors affect a person's performance in any given testing situation.

When you have finished the practice questions, you can read through the sample responses with scorer annotations in chapter 10.

THE **PRAXIS** ™

S E R I E S

TEST NAME

Mathematics: Proofs, Models, and Problems, Part 1 (0063)

Time—60 Minutes

4 Questions

NOTATIONS

(a, b)	$\{\, x : a < x < b \,\}$
$[a, b)$	$\{\, x : a \leq x < b \,\}$
$(a, b]$	$\{\, x : a < x \leq b \,\}$
$[a, b]$	$\{\, x : a \leq x \leq b \,\}$
$\gcd(m, n)$	<u>greatest common divisor</u> of two integers m and n
$\operatorname{lcm}(m, n)$	<u>least common multiple</u> of two integers m and n
$[x]$	<u>greatest integer</u> m such that $m \leq x$
$m \equiv k \pmod{n}$	m and k are <u>congruent modulo n</u> (m and k have the same remainder when divided by n, or equivalently, $m - k$ is a multiple of n)
f^{-1}	<u>inverse</u> of an invertible function f; (<u>not</u> to be read as $\frac{1}{f}$)
$\displaystyle\lim_{x \to a^+} f(x)$	<u>right-hand limit</u> of $f(x)$; limit (if it exists) of $f(x)$ as x approaches a from the right
$\displaystyle\lim_{x \to a^-} f(x)$	<u>left-hand limit</u> of $f(x)$; limit (if it exists) of $f(x)$ as x approaches a from the left
$p \to q$	statement p <u>implies</u> statement q; if p, then q
$p \leftrightarrow q$	statements p and q are <u>logically equivalent</u>; p if and only if q
$\sim p$	<u>negation</u> of the statement p; not p
$p \wedge q$	<u>conjunction</u> of statements p and q; p and q
$p \vee q$	<u>disjunction</u> of statements p and q; p or q
$p \veebar q$	<u>exclusive or</u> for statements p and q; p or q but not both
$B \leftarrow C$	(in a computer algorithm) value of the variable B is replaced by the current value of the variable C
$\varnothing$	the empty set
$x \in S$	x is an element of set S
$S \subset T$	set S is a proper subset of set T
$S \subseteq T$	either set S is a proper subset of set T or $S = T$
$\overline{S}$	complement of set S; the set of all elements not in S that are in some specified universal set
$T \setminus S$	relative complement of set S in set T, i.e., the set of all elements of T that are not elements of S
$S \cup T$	union of sets S and T
$S \cap T$	intersection of sets S and T

FORMULAS

<u>Sum</u>

$$\sin(x + y) = \sin x \cos y + \cos x \sin y$$

$$\cos(x + y) = \cos x \cos y - \sin x \sin y$$

<u>Half-Angle</u>   (sign depends on the quadrant of $\frac{\theta}{2}$)

$$\sin\frac{\theta}{2} = \pm\sqrt{\frac{1 - \cos\theta}{2}}; \qquad \cos\frac{\theta}{2} = \pm\sqrt{\frac{1 + \cos\theta}{2}}$$

<u>Range of inverse trigonometric functions</u>

$$\sin^{-1} x \ [-\pi/2, \pi/2]; \qquad \cos^{-1} x \ [0, \pi]; \qquad \tan^{-1} x \ (-\pi/2, \pi/2)$$

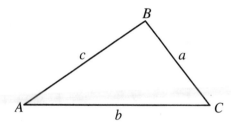

<u>Law of Sines</u> $\dfrac{\sin A}{\sin B} = \dfrac{a}{b}$

<u>Law of Cosines</u> $c^2 = a^2 + b^2 - 2ab(\cos C)$

<u>DeMoivre's Theorem</u> $(\cos\theta + i\sin\theta)^k = \cos(k\theta) + i\sin(k\theta)$

<u>Coordinate Transformation</u>

Rectangular (x, y) to polar (r, θ) $r^2 = x^2 + y^2$; $\tan\theta = \dfrac{y}{x}$ if $x \neq 0$

Polar (r, θ) to rectangular (x, y) $x = r\cos\theta$; $y = r\sin\theta$

<u>Distance from point (x_1, y_1) to line</u> $Ax + By + C = 0$ $d = \dfrac{|Ax_1 + By_1 + C|}{\sqrt{A^2 + B^2}}$

Volume

Sphere: radius r $\qquad$ $V = \frac{4}{3}\pi r^3$

Right circular cone: height h, base of radius r $\qquad$ $V = \frac{1}{3}\pi r^2 h$

Right circular cylinder: height h, base of radius r $\qquad$ $V = \pi r^2 h$

Pyramid: height h, base of area B $\qquad$ $V = \frac{1}{3}Bh$

Right prism: height h, base of area B $\qquad$ $V = Bh$

Surface Area

Sphere: radius r $\qquad$ $A = 4\pi r^2$

Lateral surface area of right circular cone: radius r, slant height s $\qquad$ $A = \pi rs$

Differentiation

$$(f(x)g(x))' = f'(x)g(x) + f(x)g'(x) \qquad (f[g(x)])' = f'[g(x)]g'(x)$$

$$\left(\frac{f(x)}{g(x)}\right)' = \frac{f'(x)g(x) - f(x)g'(x)}{(g(x))^2} \qquad \text{if } g(x) \neq 0$$

Integration by Parts $\qquad \int u \, dv = uv - \int v \, du$

DEFINITIONS

Linear Algebra

A vector $\mathbf{u}$ is a <u>linear combination</u> of the vectors $\mathbf{v}_1, \mathbf{v}_2, \mathbf{v}_3, \ldots, \mathbf{v}_n$ if there exist real numbers $a_1, a_2, a_3, \ldots, a_n$ such that $\mathbf{u} = a_1\mathbf{v}_1 + a_2\mathbf{v}_2 + a_3\mathbf{v}_3 + \ldots + a_n\mathbf{v}_n$.

The <u>linear span</u> of the vectors $\mathbf{v}_1, \mathbf{v}_2, \mathbf{v}_3, \ldots, \mathbf{v}_n$ is the set of all linear combinations of $\mathbf{v}_1, \mathbf{v}_2, \mathbf{v}_3, \ldots, \mathbf{v}_n$.

The vectors $\mathbf{v}_1, \mathbf{v}_2, \mathbf{v}_3, \ldots, \mathbf{v}_n$ are <u>linearly independent</u> if $a_1\mathbf{v}_1 + a_2\mathbf{v}_2 + a_3\mathbf{v}_3 + \ldots + a_n\mathbf{v}_n = \mathbf{0}$ (the zero vector) implies that $a_1 = a_2 = a_3 = \ldots = a_n = 0$.

The set of vectors $\mathbf{v}_1, \mathbf{v}_2, \mathbf{v}_3, \ldots, \mathbf{v}_k$ forms a <u>basis</u> for a subspace W of $\mathbf{R}^n$ if $\mathbf{v}_1, \mathbf{v}_2, \mathbf{v}_3, \ldots, \mathbf{v}_k$ are linearly independent and their linear span is equal to W.

The <u>dimension</u> of a subspace W of $\mathbf{R}^n$ is the number of vectors in a basis for W.

Discrete Mathematics

A relation $\Re$ on a set S is

    <u>reflexive</u> if $x \Re x$ for all $x \in S$

    <u>symmetric</u> if $x \Re y \Rightarrow y \Re x$ for all $x, y \in S$

    <u>transitive</u> if $(x \Re y$ and $y \Re z) \Rightarrow x \Re z$ for all $x, y, z \in S$

    <u>antisymmetric</u> if $(x \Re y$ and $y \Re z) \Rightarrow x = y$ for all $x, y, \in S$

An <u>equivalence</u> relation is a reflexive, symmetric, and transitive relation.

MATHEMATICS: PROOFS, MODELS, AND PROBLEMS, PART 1

Directions: This question is worth $16\frac{2}{3}$ percent of your score for this test. Answer all parts of the question in the space provided. The graphing calculator may be useful for some parts of the question.

Question 1

Marissa drove 500 miles and stopped only once during the trip. Her average speed for the first 250 miles of the trip was 55 miles per hour. After driving 250 miles, Marissa stopped for 1 hour to have lunch. After her stop Marissa drove at an average speed of 45 miles per hour for the rest of the trip.

(A)   What was Marissa's average speed for the 500-mile trip if the average is calculated <u>excluding</u> the time she spent having lunch? Show how you arrived at your answer.

(B)   What was Marissa's average speed for the 500-mile trip if the average is calculated <u>including</u> the time she spent having lunch? Show how you arrived at your answer.

NOTES

Begin your response to Question 1 here.

(Question 1—*Continued*)

(Question 1—*Continued*)

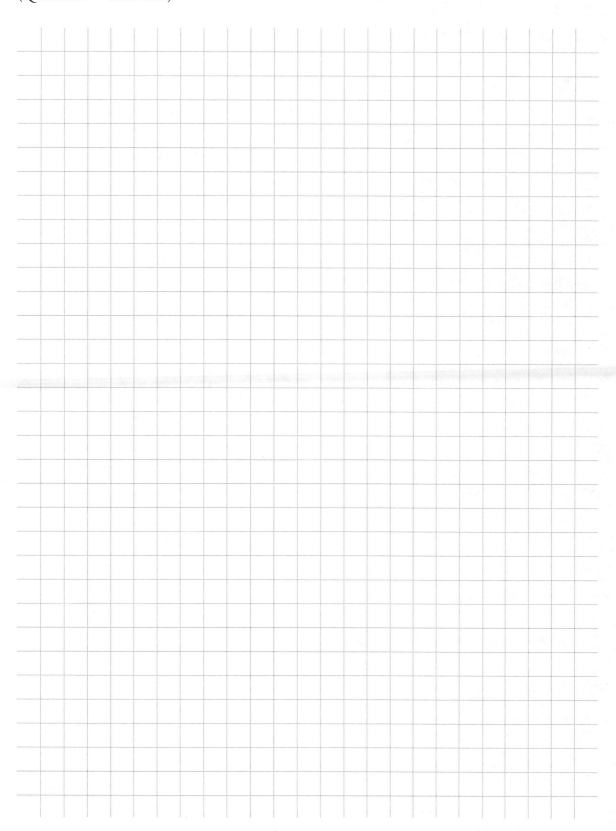

Directions: This question is worth $16\frac{2}{3}$ percent of your score for this test. Answer all parts of the question in the space provided. The graphing calculator may be useful for some parts of the question.

Question 2

(A)   Determine an equation of the function whose graph is formed by moving each point on the graph of the function $y = x^2$ up 4 units. Show how you arrived at your answer.

(B)   Determine an equation of the function whose graph is formed by moving each point on the graph of the function $y = x^2$ to the right 3 units. Show how you arrived at your answer.

(C)   Determine an equation of the function whose graph is formed by moving each point on the graph of the function $y = x^2$ to a point with the same y-coordinate and twice the x-coordinate. Show how you arrived at your answer.

NOTES

Begin your response to Question 2 here.

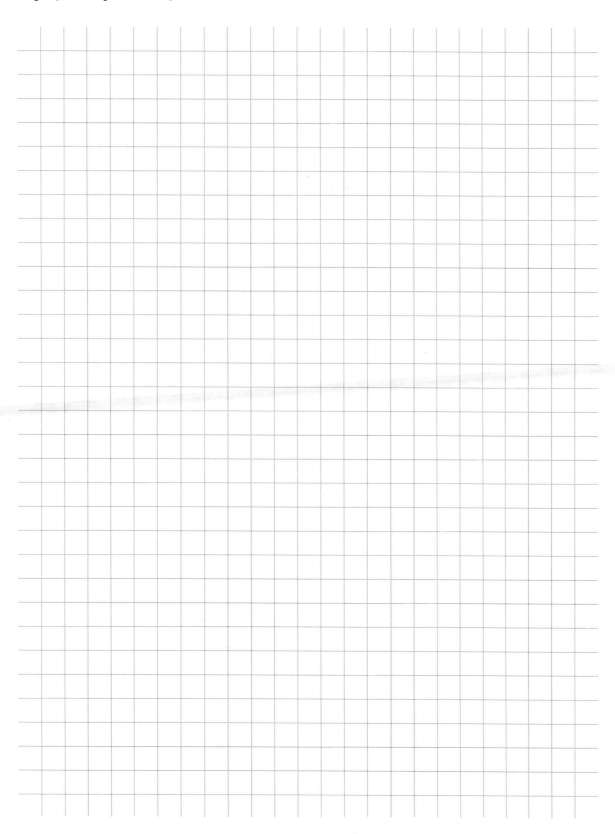

(Question 2—*Continued*)

(Question 2—*Continued*)

Directions: This question is worth $33\frac{1}{3}$ percent of your score for this test. Answer all parts of the question in the space provided. The graphing calculator may be useful for some parts of the question.

Question 3

City *A* currently has a population of 250,000, and City *B* currently has a population of 350,000. If the population of City *A* increases at a constant rate of 3% per year and the population of City *B* increases at a constant rate of 1% per year, then in approximately how many years will the population of the two cities be equal? According to this projection, what will the population of City *A* be at that time? Explain how you arrived at your answer.

NOTES

Begin your response to Question 3 here.

(Question 3—*Continued*)

(Question 3—*Continued*)

Directions: This question is worth $33\frac{1}{3}$ percent of your score for this test. Answer all parts of the question in the space provided. The graphing calculator may be useful for some parts of the question.

Question 4

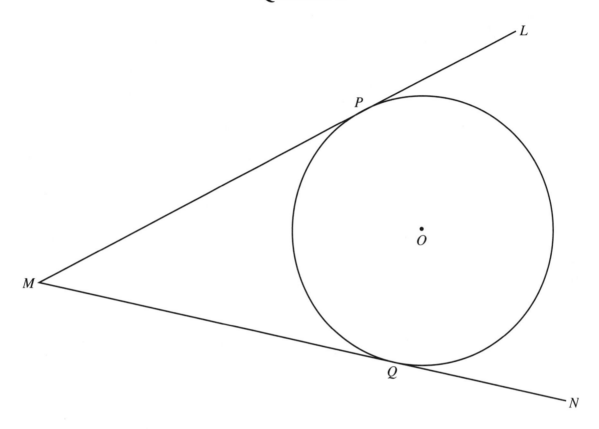

O is the center of the circle shown. *ML* and *MN* are tangent to the circle at points *P* and *Q*, respectively.

Prove that the length of line segment *MP* is equal to the length of line segment *MQ*.

NOTES

Begin your response to Question 4 here.

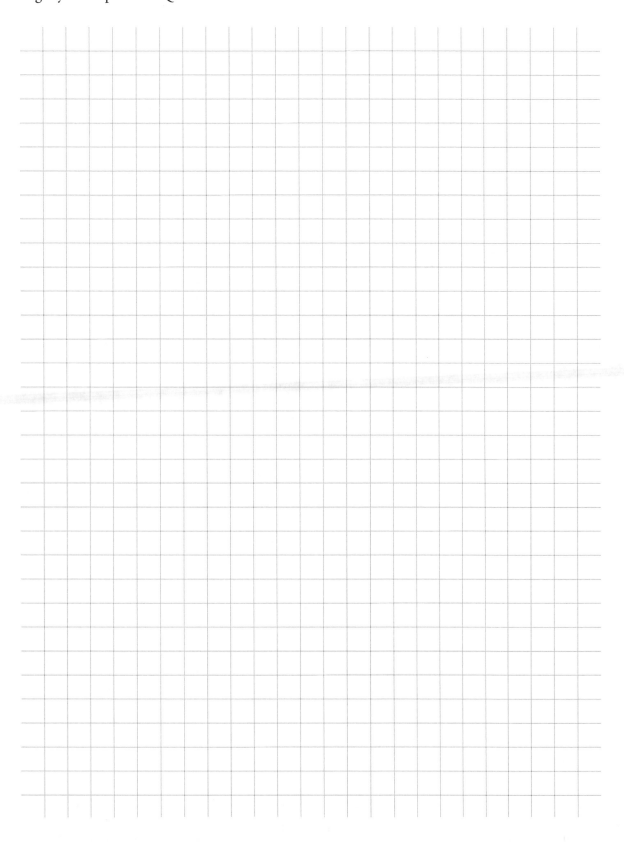

(Question 4—*Continued*)

(Question 4—*Continued*)

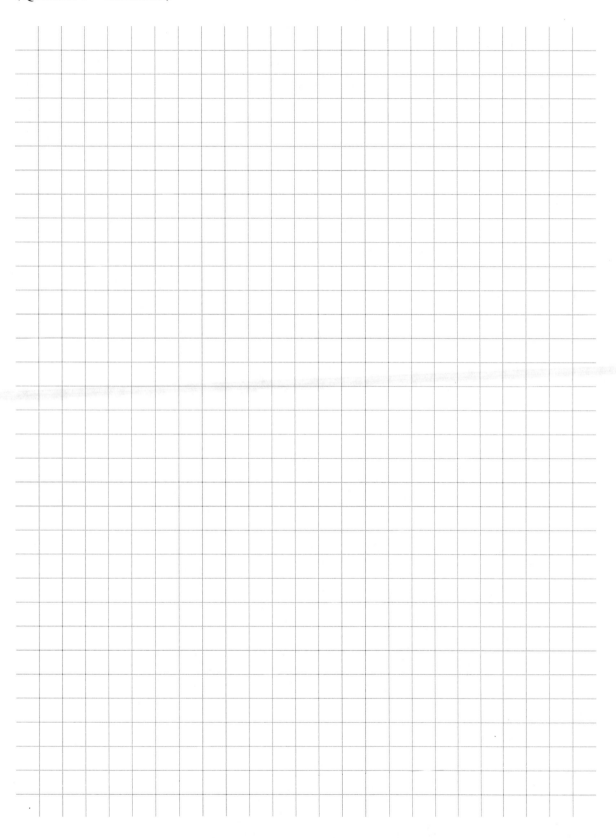

Chapter 8
Practice Test, *Mathematics: Pedagogy*

▶ ▶ ▶ ▶ ▶ ▶ ▶ ▶ ▶ ▶ ▶ ▶

Now that you have studied the content topics and have worked through strategies relating to the *Pedagogy* test, you should take the following practice test. You will probably find it helpful to simulate actual testing conditions, giving yourself 60 minutes to work on the questions. You can use the response pages provided if you wish.

Keep in mind that the test you take at an actual administration will have different questions. You should not expect your level of performance to be exactly the same as when you take the test at an actual administration, since numerous factors affect a person's performance in any given testing situation.

When you have finished the practice questions, you can read through the sample responses with scorer annotations in chapter 11.

THE **PRAXIS**™
S E R I E S

TEST NAME

Mathematics:
Pedagogy (0065)

Time—60 Minutes

3 Questions

MATHEMATICS: PEDAGOGY

Directions: Answer all parts of the following question in the space provided.

Question 1

You are teaching a unit on solving quadratic equations. You have already taught the students how to solve quadratics by taking square roots and by factoring. In your next lesson, you plan to teach the students how to solve quadratic equations by completing the square.

Design a homework assignment for your students to complete after the lesson on solving quadratic equations by completing the square. The homework assignment should consist of 5 problems that review previously taught skills and concepts while also providing practice in the newly introduced material.

Briefly explain your rationale for including the skills and concepts that the problems illustrate.

NOTES

Begin your response to Question 1 here.

(Question 1—*Continued*)

(Question 1—*Continued*)

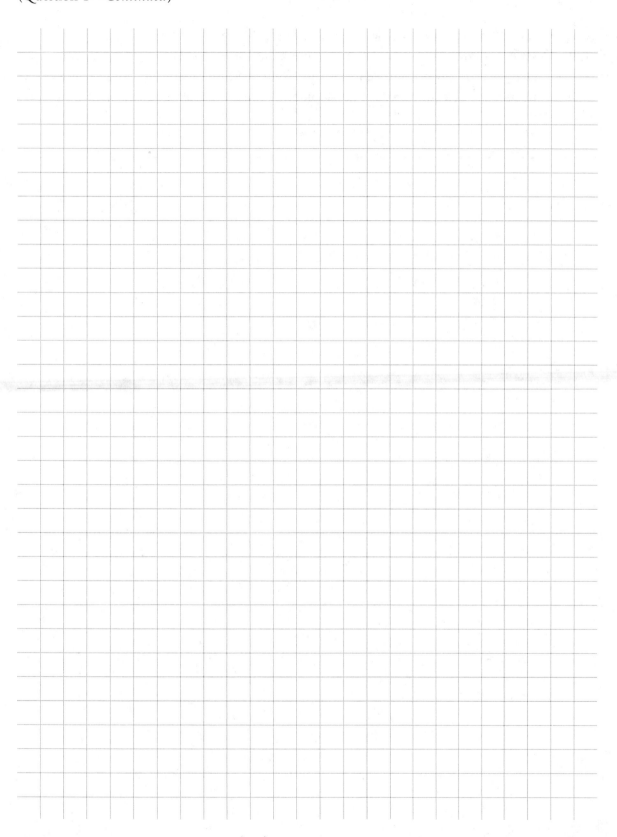

Question 2

A small group of students in your seventh-grade math class is unable to determine whether two fractions are equivalent. Describe a strategy, using pictures or manipulatives, that you could use to help foster the students' conceptual understanding of equivalent fractions. Your strategy should stress understanding what it means for fractions to be equivalent and developing the ability to determine whether fractions are equivalent.

NOTES

Begin your response to Question 2 here.

(Question 2—*Continued*)

(Question 2—*Continued*)

Question 3

Students in geometry frequently confuse the concepts of "equilateral" and "regular." For a triangle, these mean the same thing, but a polygon with more than 3 sides can be equilateral without being regular. Describe an investigation you would have your students make to discover the difference between "regular" and "equilateral." Your investigation may involve the use of any type of manipulative or any software package.

NOTES

Begin your response to Question 3 here.

(Question 3—*Continued*)

(Question 3—*Continued*)

Chapter 9

Explanations of Answers for the
Mathematics: Content Knowledge Test

▶ ▶ ▶ ▶ ▶ ▶ ▶ ▶ ▶ ▶ ▶ ▶

Scoring Your Practice Test

To score your *Mathematics: Content Knowledge* practice test:

- Count the number of questions you answered correctly. The correct answers are in Table 1.
- Use Table 2 to find the scaled score corresponding to the raw score (number of questions answered correctly). You can compare your scaled score to the passing score required by your state or institution. (Passing state scores are available on the Praxis Web site at www.ets.org/praxis.)

Right Answers and Explanations for the Practice Test

Table 1—Answers and Content Categories for the Practice Test Questions

Question Number	Correct Answer	Content Category	Question Number	Correct Answer	Content Category
1	B	Discrete Mathematics	26	A	Arithmetic and Basic Algebra
2	B	Mathematical Reasoning and Modeling	27	A	Functions and Their Graphs
3	C	Discrete Mathematics	28	C	Arithmetic and Basic Algebra
4	C	Functions and Their Graphs	29	B	Computer Science
5	C	Probability and Statistics	30	A	Geometry
6	A	Arithmetic and Basic Algebra	31	C	Arithmetic and Basic Algebra
7	A	Mathematical Reasoning and Modeling	32	B	Calculus
8	D	Analytic Geometry	33	C	Geometry
9	B	Geometry	34	C	Linear Algebra
10	C	Analytic Geometry	35	B	Mathematical Reasoning and Modeling
11	D	Trigonometry	36	A	Calculus
12	D	Probability and Statistics	37	C	Arithmetic and Basic Algebra
13	A	Linear Algebra	38	A	Analytic Geometry
14	B	Computer Science	39	A	Analytic Geometry
15	D	Calculus	40	D	Trigonometry
16	C	Mathematical Reasoning and Modeling	41	A	Probability and Statistics
17	C	Discrete Mathematics	42	B	Functions and Their Graphs
18	C	Geometry	43	A	Linear Algebra
19	B	Mathematical Reasoning and Modeling	44	C	Mathematical Reasoning and Modeling
20	B	Probability and Statistics	45	C	Functions and Their Graphs
21	B	Calculus	46	D	Geometry
22	C	Functions and Their Graphs	47	C	Arithmetic and Basic Algebra
23	D	Geometry	48	B	Functions and Their Graphs
24	D	Probability and Statistics	49	B	Probability and Statistics
25	D	Calculus	50	D	Functions and Their Graphs

Table 2—Conversion Table

Raw Score	Scaled Score	Raw Score	Scaled Score	Raw Score	Scaled Score
0	100	17	107	34	151
1	100	18	110	35	154
2	100	19	113	36	156
3	100	20	116	37	159
4	100	21	118	38	161
5	100	22	121	39	164
6	100	23	124	40	167
7	100	24	126	41	170
8	100	25	129	42	174
9	100	26	131	43	178
10	100	27	134	44	182
11	100	28	136	45	186
12	100	29	139	46	190
13	100	30	141	47	194
14	100	31	144	48	198
15	102	32	146	49	200
16	105	33	149	50	200

Estimating an actual score from your practice score

When you take the *Mathematics: Content Knowledge* test at an actual administration, the questions you will be presented with will be similar to the questions in this practice test, but they will not be identical. Because of the differences in questions, the test that you actually take may be slightly more or less difficult.[1] Therefore, you should not expect to get exactly the same raw score that you achieved on this practice test.

---

[1] There is a statistical adjustment for difficulty (using a process called "equating") that makes it possible to give the same interpretation to identical scaled scores on different editions of the same test.

Explanations of right answers

1. One way to get started on this problem is to draw three lines arranged in several different ways on a blank sheet of paper.

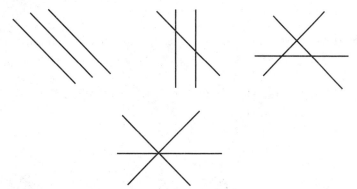

   Examining these figures helps us determine all of the possible arrangements of the three lines. The possible arrangements are

   - all three lines are parallel,
   - exactly two of the lines are parallel,
   - the three lines have a common point of intersection, and
   - no pair of the lines are parallel and the three lines do not have a common point of intersection.

   By looking at the figures of each of these cases, we see that the plane is divided into the most regions when no pair of the lines are parallel and the three lines do not have a common point of intersection.

   Since in this case the plane is divided into 7 disjoint regions, the correct answer is B.

2.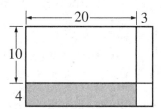

   Based on the figure and the fact that the shaded region is a rectangle, we can see that the sides of the shaded rectangle have lengths 4 and 20, respectively, and the area of the region is 4×20.

   Therefore, the correct answer is B.

3. In order to count all of the different sandwiches José can choose, it may be easiest to break the problem down into two different cases and then count all of the possibilities for each case.

   Case I: José chooses a sandwich with only 1 filling (i.e., either one type of cheese or one type of meat).

   In this case, José can choose either of 2 types of bread and any 1 of 7 different types of filling. So the total number of different sandwiches in Case I is $2 \times 7 = 14$.

   Case I   2×7 = 14

+ Case II   2 × 3×4 = 24
+ ___
38

Case II: José chooses a sandwich with 2 fillings (i.e., one type of cheese and one type of meat). Note that, according to the statement of the problem, 1 of the 2 fillings must be one type of cheese and the other must be one type of meat.

In this case, José can choose either of 2 types of bread, any one of the 3 cheeses, and any one of the 4 meats. So the total number of sandwiches in Case II is $2 \times 3 \times 4 = 24$.

Cases I and II cover all the possible sandwiches that José can choose, so the total number of different sandwiches that he can choose is $14 + 24 = 38$.

Therefore, the correct answer is C.

4. The graph of the function $y = f(x + 1)$ is the same as the graph of $y = f(x)$ shifted left by 1 unit. Also, the graph of $y = f(x + 1) - 2$ is the same as the graph of $y = f(x + 1)$ shifted down by 2 units.

Therefore, the graph of $y = f(x + 1) - 2$ is the same as the graph of $y = f(x)$ shifted left by 1 unit and shifted down by 2 units.

Therefore, the correct answer is C.

Suppose that you remember that the graph of the function $y = f(x + 1) - 2$ is the same as the graph of $y = f(x)$ but you do not remember how to shift the graph. One strategy for solving the problem is calculate the value of $y = f(x + 1) - 2$ for a particular value of x and see where it "came from" on the graph of $y = f(x)$.

Selecting $x = 0$ (since it is easy to work with), you can see that $y = f(0 + 1) - 2 = f(1) - 2$. This is the value of $y = f(x)$ at $x = 1$ shifted to the left by 1 unit to get from $x = 1$ to $x = 0$ and down by 2 units.

Suppose that you do not remember how the graph of $y = f(x + 1) - 2$ relates to the graph of $y = f(x)$. A good strategy for solving the problem is to look at the options to see if they give you any information about the answer to the problem.

Notice that every option contains the graph of a function that is the graph of $y = f(x)$ moved to a different position in the plane. You can, therefore, conclude that the graph of $y = f(x + 1) - 2$ is the graph of $y = f(x)$ moved to a different position in the plane. For any function $y = f(x)$, the movement is the same. Therefore, you can select any function $y = f(x)$ and use the graphing calculator to graph the functions $y = f(x)$ and $y = f(x + 1) - 2$ and see the relationship between them.

How do you select the function to graph? It is useful to select a function that you know how to express algebraically and that has a graph that will be easy to compare with the shifted graph. One such function is $y = x^2$.

Suppose that you graph $y = x^2$ and $y = (x + 1)^2 - 2$ in the viewing window $[-10,10] \times [-10,10]$. Your calculator would show the following.

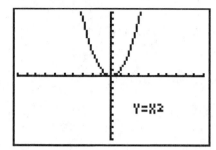 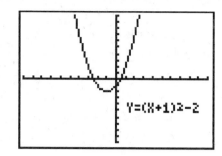

Notice that the graph of $y = x^2$ has been shifted down and to the left to give the graph of $y = f(x+1)^2 - 2$. Therefore, shifting the graph in the question down and to the left gives the graph in C.

5. In order to find the mean number of days per student it took to complete the assignment, we should first find the total number of days all of the students combined spent on the assignment and then divide by the total number of students (50).

   5 students used 1 day each for the assignment, so those students used a total of 5 days.
   10 students used 2 days each on the assignment, so those students used a total of 20 days.
   15 students used 3 days each on the assignment, so those students used a total of 45 days.
   15 students used 4 days each on the assignment, so those students used a total of 60 days.
   5 students used 5 days each on the assignment, so those students used a total of 25 days.

   Adding each of these numbers, we see that a total of $5 + 20 + 45 + 60 + 25 = 155$ days were used by all students on the assignment. Therefore, the mean number of days per student taken to complete the assignment is $\frac{155}{50} = 3.1$.

   Therefore, the correct answer is C.

6. For absolute value problems, it is often useful to recall that $|a - b|$ represents the distance from point a to point b on the number line. Since $a < b$, the distance from point a to point b must be greater than 0, so that $|a - b| > 0$.

   Therefore, the correct answer is A.

   You can see that B, C, and D are not necessarily true by thinking of a counterexample for each one.

   (B) A counterexample for B can be obtained by noting that the statement $|a| < |b|$ can be interpreted on a number line as

   a is closer to 0 than b is.

   Clearly $-3 < 1$, but 1 is closer to 0 than -3 is.

(C) A counterexample for C can be obtained by noting that the statement $|a| = |b|$ can be interpreted on the number line as

$$a \text{ and } b \text{ are the same distance from } 0.$$

Clearly, $-1 < 1$ but 1 and -1 are the same distance from 0.

(D) A counterexample for D can be obtained by noting that the statement $|b| - |a| < |b - a|$ can be interpreted on the number line as

[the distance between b and 0] minus [the distance between a and 0] is less than [the distance between b and a].

The two quantities are equal if a and b are on the same side of 0 on the number line, as shown on the figure below.

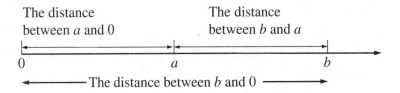

Clearly $1 < 2$, but it is not true that $|2| - |1| < |2 - 1|$.

7. (A) The <u>integer</u> multiples of $\frac{1}{2}$ between 0 and 1,000 are

$$1 \times \frac{1}{2}, \ 2 \times \frac{1}{2}, \ 3 \times \frac{1}{2}, \ \dots, \ 1{,}998 \times \frac{1}{2}, \ 1{,}999 \times \frac{1}{2}.$$

There are clearly only a finite number (1,999) of these.

Therefore, the correct answer is A.

(B) To see that the set in B is infinite, one can consider the numbers

$$7.1, \ 7.01, \ 7.001, \ 7.0001, \ \dots,$$

where the list continues indefinitely. This is an infinite set of rational numbers that are between 7 and 8.

(C) The same numbers as those shown for B show that the set in C is infinite, since each of the numbers in that list is rational and has a terminating decimal expansion.

(D) To see that there are infinitely many irrational numbers, you can consider the numbers

$$\sqrt{2}, \ 2 \times \sqrt{2}, \ 3 \times \sqrt{2}, \ 4 \times \sqrt{2}, \ \dots;$$

that is, all positive integer multiples of $\sqrt{2}$. Since $\sqrt{2}$ is irrational, each number in the list is irrational, and there are infinitely many numbers in the list.

8. By graphing a parabola that is concave up or "opens up," we can see that the axis of symmetry for the parabola (i.e., the line around which the parabola is symmetric) is the vertical line that passes through the vertex of the parabola. In addition, we can see that the vertex is at the lowest point on the parabola. To solve the problem, we need to find the x-coordinate of the vertex. Using the graphing calculator, we can see that the parabola has a minimum at $x = 2.75$.

Therefore, the answer is D.

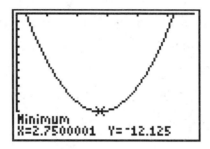

Another way to solve the problem is to notice that parabolas with equations of the given form have (i) an axis of symmetry parallel to the y-axis passing through the vertex of the parabola and (ii) an absolute minimum at the vertex. Thus, the equation of the axis of symmetry is $x = c$, where c is the x-coordinate of the vertex. To find c, you can use calculus to find the value of x that minimizes $y = 2x^2 - 11x + 3$.

If $y = 2x^2 - 11x + 3$, then $\dfrac{dy}{dx} = 4x - 11$. The minimum of the function occurs at the value of x where the derivative is equal to 0, and $\dfrac{dy}{dx} = 4x - 11 = 0$ only when $x = \dfrac{11}{4} = 2.75$.

Alternatively, once you have realized that the axis of symmetry passes through the vertex (b, c) the equation of the parabola can be expressed in the form $y = a(x - b)^2 + c$. So the problem can be solved by putting $y = 2x^2 - 11x + 3$ into this form.

$$y = 2x^2 - 11x + 3$$
$$y = 2\left(x^2 - \frac{11}{2}x + \frac{121}{16}\right) - \frac{121}{8} + 3$$
$$y = 2\left(x - \frac{11}{4}\right)^2 - \frac{97}{8}$$

Thus, $b = \dfrac{11}{4} = 2.75$ and the axis of symmetry is $x = 2.75$.

9.

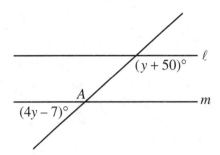

Label angle A as shown in the figure above. Since lines ℓ and m are parallel, angle A and the angle measuring $(y + 50)$ degrees are alternate interior angles and so are congruent. Therefore, the measure of angle A is equal to $(y + 50)$ degrees.

Since angle A and the angle measuring $4y - 7$ degrees are supplementary (they combine to form a straight line), $(4y - 7) + (y + 50) = 180$.

When we solve for y: $\quad 5y + 43 = 180$
$$5y = 137$$
$$y = \frac{137}{5}$$

Therefore, the correct answer is B.

Note: Here are some useful facts for problems involving angles and parallel lines.

- Alternate interior angles are congruent.
- Opposite angles are congruent.
- The degree measures of consecutive interior angles sum to 180 degrees.
- The degree measures of supplementary angles sum to 180 degrees.

10.

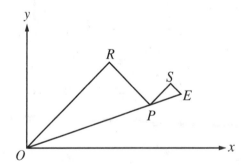

When starting this problem, you may find it useful to draw two hypotenuses (segments OP and PE) to form two right triangles. Because the legs of the large triangle are each 4 times as long as the corresponding legs of the smaller triangle, these two triangles are similar. Because the two triangles are similar and are right triangles, angle measures can be added to the figure as follows.

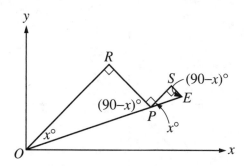

You can see from the figure that the degree measure of angle *OPE* is 180 and line segment *OP* and line segment *PE* lie on the same line.

Since segments *OP* and *PE* are on the same line, the distance from point *E* to the origin will be the sum of the lengths of the hypotenuses. Using the Pythagorean theorem, we see that the lengths are as follows.

$$\text{Length of } OP = \sqrt{8^2 + 4^2} = \sqrt{64+16} = \sqrt{80} = 4\sqrt{5}$$
$$\text{Length of } PE = \sqrt{2^2 + 1^2} = \sqrt{5}$$

Therefore, the distance from point *E* to the origin is $4\sqrt{5} + \sqrt{5} = 5\sqrt{5}$, and the correct answer is C.

11. One way to approach this problem is to use the fact that $\tan x = \dfrac{\sin x}{\cos x}$, along with the trigonometric identity $\sin^2 x + \cos^2 x = 1$.

$$\frac{\sin x}{\cos x} = \frac{2}{3} \Rightarrow \sin x = \frac{2}{3}\cos x.$$

Substituting for sin *x* in the identity, we have

$$\frac{4}{9}\cos^2 x + \cos^2 x = 1 \Rightarrow \frac{13}{9}\cos^2 x = 1 \Rightarrow \cos^2 x = \frac{9}{13}.$$

Solving for cos *x* gives $\cos x = \pm\dfrac{3}{\sqrt{13}}$, and since $\pi < x < \dfrac{3\pi}{2}$, cos *x* must be negative and so $\cos x = -\dfrac{3}{\sqrt{13}}$. The correct answer is D.

Another way to approach this problem is to use the coordinate plane and the definition of the trigonometric functions of angles in a right triangle to angles outside the range $0 < x < \dfrac{\pi}{2}$. Since $\pi < x < \dfrac{3\pi}{2}$, we get the following graph.

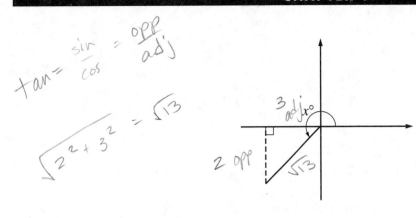

The hypotenuse of the reference triangle is $\sqrt{(-2)^2 + (-3)^2} = \sqrt{13}$, and $\cos x = -\dfrac{3}{\sqrt{13}}$.

12. Since each standard deviation equals 8, then 2.5 standard deviations are equal to $2.5 \times 8 = 20$, and the value 2.5 standard deviations above the mean is $60 + 20 = 80$.

Therefore, the correct answer is D.

13. If we write down a general 2×2 matrix as $A = \begin{pmatrix} a & b \\ c & d \end{pmatrix}$, then

$$A\begin{pmatrix} x \\ y \end{pmatrix} = \begin{pmatrix} a & b \\ c & d \end{pmatrix}\begin{pmatrix} x \\ y \end{pmatrix} = \begin{pmatrix} ax + by \\ cx + dy \end{pmatrix}.$$

Since this equals $3\begin{pmatrix} x \\ y \end{pmatrix} = \begin{pmatrix} 3x \\ 3y \end{pmatrix}$, the following equations must hold for any x and y.

$$ax + by = 3x$$
$$cx + dy = 3y$$

The only way this can be true for every x and y is if $a = 3$, $b = 0$, $c = 0$, and $d = 3$.

Therefore, $A = \begin{pmatrix} 3 & 0 \\ 0 & 3 \end{pmatrix}$ and the correct answer is A.

As an alternative solution, you could note that multiplying any vector $\begin{pmatrix} x \\ y \end{pmatrix}$ by A has the same effect as multiplying that vector by 3. This is true only when A is 3 times the identity matrix $I = \begin{pmatrix} 1 & 0 \\ 0 & 1 \end{pmatrix}$. Therefore, $A = 3I = \begin{pmatrix} 3 & 0 \\ 0 & 3 \end{pmatrix}$.

14. In STEP 1, X is set equal to 100 and Y is set equal to 10.

In STEP 2, since $X \geq 100$, the "if" line is true and the "then" line is executed. The value of X is replaced by $X + 1 = 100 + 1 = 101$. After STEP 2 is executed, $X = 101$ and $Y = 10$.

In STEP 3, $X > Y$ but Y is not less than 10, so the "if" line does <u>not</u> execute. Instead, the "else" line executes and the value of X is replaced by $X - Y = 101 - 10 = 91$.

After STEP 3 has executed, $X = 91$ and $Y = 10$.

In STEP 4, the current value of X is printed. This value is 91.

Therefore, the correct answer is B.

15. For curves above the x-axis, integration represents the area bounded by the curve and the x-axis. For curves below the x-axis, integration represents the <u>negative</u> of the area bounded by the curve and the x-axis.

The region below the curve and above the x-axis for $-2 < x < 1$ is very similar in shape and size to the region above the curve and below the x-axis for $1 < x < 4$. Since the areas of these two regions are approximately equal while one is above the x-axis and the other below the x-axis, the integral over the entire interval $-2 < x < 4$ will be approximately 0, i.e., $\int_{-2}^{4} f(x)dx \approx 0$.

Therefore, the correct answer is D.

Notice that in this approach, since every area above the x-axis was "canceled" by an area below the x-axis, no calculation had to be done to solve the problem. In a problem involving integration with areas above and below the x-axis, only areas that do not "cancel" have to be calculated to evaluate the integral.

An alternative approach is to estimate the area above the x-axis and the area below the y-axis by counting boxes and fractions of boxes.

16. This is as much a counting problem as a geometry problem. We need to find all points (x, y) with integer coordinates that satisfy $x^2 + y^2 \leq 9$. To generate a complete list of those points, it is useful to start with the points that have x-coordinate equal to 0 and then move on to points having nonzero x-coordinates.

$x = 0$:	$(0, 0)$	$(0, 1)$	$(0, -1)$	$(0, 2)$	$(0, -2)$	$(0, 3)$ $(0, -3)$	7 points
$x = 1$:	$(1, 0)$	$(1, 1)$	$(1, -1)$	$(1, 2)$	$(1, -2)$		5 points
$x = -1$:	$(-1, 0)$	$(-1, 1)$	$(-1, -1)$	$(-1, 2)$	$(-1, -2)$		5 points
$x = 2$:	$(2, 0)$	$(2, 1)$	$(2, -1)$	$(2, 2)$	$(2, -2)$		5 points
$x = -2$:	$(-2, 0)$	$(-2, 1)$	$(-2, -1)$	$(-2, 2)$	$(-2, -2)$		5 points
$x = 3$:	$(3, 0)$						1 point
$x = -3$:	$(-3, 0)$						1 point

Adding up the number of points for each different value of the *x*-coordinate, we see that there are 29 points with integer coordinates inside or on the circle.

Therefore, the correct answer is C.

17. You can start this problem by calling the teams Team A, Team B, Team C, Team D, and Team E.

Team A will play each of the other teams twice for a total of 8 games.

Team B will play each of the other teams twice also, but Team B's 2 games with Team A have already been counted above, so we need only count Team B's games with Team C, Team D, and Team E: 6 games.

The only games of Team C that are not counted yet are those with Team D and Team E: 4 games.

All of Team D's games have been counted except for the two games with Team E: 2 games.

All of Team E's games have already been counted: 0 games.

Adding up all of the games, we see that there will be a total of $8 + 6 + 4 + 2 + 0 = 20$ games played among the five teams.

Therefore, the answer is C.

Another way to approach this problem is to write the letter of each team and draw a line segment connecting each pair of teams, as shown in the figure below.

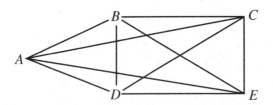

Each line segment represents 2 games, and there are 10 line segments in the figure. Thus 20 games are played during the season.

As an alternative solution, note that there are $\binom{5}{2} = 10$ different possible pairings among the 5 teams and that each pairing will play 2 games, so there will be a total of $2 \times 10 = 20$ games.

18. Since line segment *RS* is a diameter of the circle, $\angle RTS$ must have a measure of 90 degrees. Since $\angle SRT$ measures 50 degrees, the measure of $\angle RST$ must be 40 degrees.

Arc *RT* is subtended by $\angle RST$, so its degree measure must be $2 \times 40 = 80$ degrees.

Therefore, the correct answer is C.

19. In order to see the behavior of the function over time, it is useful to graph the function. Notice that at time $t = 0$, the value of the function is $\frac{10,000}{10+50}$, or $166\frac{2}{3}$. So a good viewing window to start with might be $[0,20] \times [0,1,000]$. Graphing the function in this viewing window shows that the function grows rapidly at first and then levels off. Zooming shows that the function does stay level. Therefore, the correct answer is B.

As an alternative solution, one could note that as $t \to \infty$, the denominator of $f(t)$ approaches 10, so that the function approaches 1,000 for the larger values of t. This eliminates A as a possible answer and suggests that B is the answer.

Also, note that $f'(t) = \frac{250,000e^{-0.5t}}{\left(10+50e^{-0.5t}\right)^2}$. This derivative is not negative for any value of t, so the function never decreases. Therefore, C and D are eliminated as possible answers.

Since A, C, and D have been eliminated as possible answers, the correct answer must be B.

As another alternative solution, one could use the "table" function on the graphing calculator to find the value of $f(t)$ for a number of values of t. Doing this for integer t from 0 to 20, for example, confirms that the function increases for several years, then levels off. The table (with values of $f(t)$ rounded to two decimal places) is shown below. Note that the table also confirms the conclusion above that the function approaches 1,000 as t increases.

t	$f(t)$	t	$f(t)$
0	166.67	11	979.98
1	247.98	12	987.76
2	352.19	13	992.54
3	472.67	14	995.46
4	596.42	15	997.24
5	709.01	16	998.33
6	800.68	17	998.98
7	868.82	18	999.38
8	916.10	19	999.63
9	947.38	20	999.77
10	967.41		

20. According to the problem, Jane's course grade will be calculated using the following formula.

Grade = 0.2 (score on test 1) + 0.2 (score on test 2) + 0.2 (score on test 3)
+ 0.2 (score on paper) + 0.05 (score on quiz 1) + 0.05 (score on quiz 2)
+ 0.1 (class participation score)

Substituting the numbers in the problem shows that Jane needs a class participation score C that satisfies the following equation.

$$90 = 0.2(93) + 0.2(82) + 0.2(89) + 0.2(95) + 0.05(86) + 0.05(96) + 0.1(C)$$

or

$$90 = 0.2(359) + 0.05(182) + 0.1(C)$$

Solving for C gives $C = \dfrac{90 - 0.2(359) - 0.05(182)}{0.1} = 91$.

Therefore, the correct answer is B.

21. Substituting $x = 3$ into the expression yields 0 in the both the numerator and the denominator. You can conclude that $(x - 3)$ is a factor of the expression in the numerator, and then factoring out $(x - 3)$ in the numerator yields $\lim\limits_{x \to 3} \dfrac{(x-3)(x^2+2)}{x-3} = \lim\limits_{x \to 3}(x^2+2) = 11$.

Therefore, the correct answer is B.

As an alternative solution, note that we can apply L'Hôpital's rule from calculus to find the limit, since substituting $x = 3$ into the expression yields a limit of the form $\dfrac{0}{0}$. Taking the derivatives of the numerator and denominator yields $\lim\limits_{x \to 3} \dfrac{x^3 - 3x^2 + 2x - 6}{x - 3} = \lim\limits_{x \to 3} \dfrac{3x^2 - 6x + 2}{1} = \dfrac{11}{1} = 11$.

22. An earthquake's rating R on the Richter scale can be represented as the logarithm base 10 of its magnitude M. In other words, $R = \log_{10} M$.

If the earthquakes are called earthquake A and earthquake B, then the information we are given can be written as $R_A = 6.8$ and $R_B = 6.6$.

$$0.2 = R_A - R_B$$
$$0.2 = \log_{10} M_A - \log_{10} M_B$$
$$0.2 = \log_{10} \frac{M_A}{M_B}$$
$$10^{0.2} = \frac{M_A}{M_B}$$

Since $10^{0.2} \approx 1.58$, the ratio of the magnitudes of the two earthquakes is approximately 158%.

Therefore, the correct answer is C.

Alternatively, since the logarithms and exponents are related, an earthquake measuring 7 on the Richter scale has a magnitude of $c \cdot 10^x$ and an earthquake measuring 6 has a magnitude of $c \cdot 10^{x-1}$.

Thus, earthquakes measuring 6.8 and 6.6 have magnitudes $c \cdot 10^{x-0.2}$ and $c \cdot 10^{x-0.4}$, respectively.

To find the ratio of the magnitudes, evaluate the expression $\dfrac{c \cdot 10^{x-0.2}}{c \cdot 10^{x-0.4}}$. This expression reduces to $10^{0.2}$, or 1.58, and the earthquake measuring 6.8 on the Richter scale is 158% of the magnitude of an earthquake measuring 6.6.

23. The surface area S of a sphere is given in terms of its radius r by the formula $S = 4\pi r^2$. *(This formula and the formula for the volume of a sphere will be contained on the information pages of your test booklet.)* So a sphere with surface area equal to 1,500 square inches will have a radius

$$r = \sqrt{\frac{1,500}{4\pi}} \approx 10.93 \text{ inches.}$$

For a sphere with radius r, the volume V of the sphere is given by $V = \frac{4}{3}\pi r^3$, so for a sphere with radius $r \approx 10.93$ inches, the volume of the sphere is given by $V = \frac{4}{3}\pi (10.93)^3 \approx 5,470$ cubic inches.

Therefore, the correct answer is D.

24.

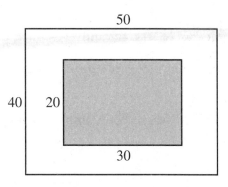

Since the position of the rabbit is uniformly random throughout the field, the probability that the rabbit will be in any region of the field is equal to the ratio of the area of that region to the area of the entire field.

The area of the entire field is $40 \times 50 = 2,000$ square feet.

The shaded region in the figure above represents the set of points in the field that are 10 or more feet away from the fence. As you can see from the figure, the shaded region is a rectangle that is 30 feet long by 20 feet wide. The area of this region is $20 \times 30 = 600$ square feet. So the probability that the rabbit will be in this region at any given time is $\dfrac{600}{2,000} = 0.30$.

Therefore, the correct answer is D.

25. A function f is strictly increasing if, for all real numbers a and b, $a < b$ implies $f(a) < f(b)$. All of the options deal with whether the derivatives of strictly increasing functions are themselves always positive and strictly increasing.

A good way to begin to determine this is to look at the derivatives of some strictly increasing functions and see whether they are always positive and/or strictly increasing. You can stop looking when you have found that you have eliminated three of the options as answers.

Some elementary examples of strictly increasing, differentiable functions are $f(x) = x$, $g(x) = x^3$, and $h(x) = e^x$.

The derivative of the function $f(x) = x$ is the function $f'(x) = 1$, which is positive but *not* strictly increasing. Thus the answer cannot be A or C.

The derivative of the function $g(x) = x^3$ is the function $g'(x) = 3x^2$, which is *not* positive when $x = 0$ and is decreasing when $x < 0$. Thus the answer cannot be A, B, or C. Therefore, the answer must be D.

26. One way to show that a set is *not* closed under division is to find members a and b of the set such that $\frac{a}{b}$ is not in the set.

If from the set of nonzero integers, we select $a = 2$ and $b = 3$, then we can see that $\frac{a}{b} = \frac{2}{3}$, which is not in the set of nonzero integers. Thus the set of nonzero integers is not closed under division. Therefore the correct answer is A.

All of the other sets are closed under division. For example, if a and b are nonzero real numbers, then $\frac{a}{b}$ is a nonzero real number.

27. To solve this problem algebraically, it is useful to find a formula for $h(f(g(x)))$. Notice that $h(f(g(x)))$ is a composition of the functions g, f, and h. In order to find the formula, we perform the composition step-by-step as follows.

$$h(f(g(x))) = h(f(x^2 + 1)) = h(4(x^2 + 1)) = \frac{1}{4(x^2 + 1)}$$

The answer will be the value of x for which $\frac{1}{4(x^2 + 1)} = \frac{1}{4}$. Solving this equation for x yields $x = 0$.

Therefore the correct answer is A.

Note: This answer could be verified by checking to see that $h(f(g(0))) = \frac{1}{4}$.

$$g(0) = 1 \implies f(g(0)) = f(1) = 4 \implies h(f(g(0))) = h(4) = \frac{1}{4}.$$

Alternatively, use the graphing calculator to graph the functions Y_3 and Y_4 defined as follows.

$$Y_1 = x^2 + 1$$
$$Y_2 = 4Y_1$$
$$Y_3 = \frac{1}{Y_2}$$
$$Y_4 = \frac{1}{4}$$
$$Y_3(x) = h(f(g(x)))$$

So the x-coordinates of the points of intersection of the graphs of Y_3 and Y_4 are the values of x that satisfy the equation $h(f(g(x))) = \frac{1}{4}$. (If the graphs did not intersect, the answer would have to be D, since it is the only option that is not a real number.) However, the x-coordinate of the intersection of Y_3 and Y_4 is 0, and the answer is A.

28. The repeating pattern of the decimal expansion of a fraction with a denominator of 7 can be shown to have at most 6 digits. The calculator can be used to evaluate $\frac{3}{7}$ as a decimal. The first 7 decimal places are 0.4285714. So it can be concluded that the pattern is 6 digits long and begins repeating in the seventh place after the decimal point. Therefore, $\frac{3}{7} = 0.\overline{428571}$.

    This pattern in decimal places 1 through 6 repeats in decimal places 7 through 12, 13 through 18, and begins again in the 19th decimal place. Thus the digit in the 19th decimal place is 4, and the correct answer is C.

29. STEP 1: 6735.8291 is read as the value of N.

    STEP 2: Several smaller steps are executed in STEP 2.
    First, N is multiplied by 100 to yield 673582.91.
    Second, 0.5 is added to that number to yield 673583.41.
    Third, the function INT selects the greatest integer less than or equal to 673583.41, which is 673583.
    Fourth, this number is divided by 100 to yield $X = 6735.83$.

    STEP 3: The value of X is printed.

    Therefore, the correct answer is B.

30. To easily remember the lengths of CD and AD, it is useful to put the lengths on the figure. Noting that the three triangles are all right triangles and setting the measure of angle C to be x degrees,

you can label the degree measures of all angles in the figure. The figure, with this information, looks like this.

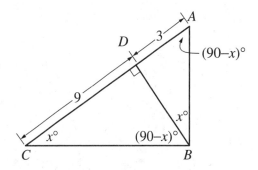

From the labeled figure, it is clear that $\triangle ABC$, $\triangle ADB$, and $\triangle BDC$ are similar triangles. Since $\triangle ABD \sim \triangle BDC$ it follows that $\dfrac{AD}{BD} = \dfrac{BD}{CD}$, which means that $\dfrac{3}{BD} = \dfrac{BD}{9}$. Solving this equation for BD yields $BD = 3\sqrt{3}$.

Now that AD and BD are known, it can be determined from the Pythagorean theorem that $AB = 6$. Similarly, the Pythagorean theorem can be used on $\triangle BDC$ to see that $BC = \sqrt{108} = 6\sqrt{3}$.

Therefore, $BC + AB + BD = 6\sqrt{3} + 6 + 3\sqrt{3} = 6 + 9\sqrt{3}$ and the correct answer is A.

Note: There are many other ways to come up with each of these lengths, using rules for similar triangles and the Pythagorean theorem.

Alternatively, beginning as in the solution above, we label the triangle.

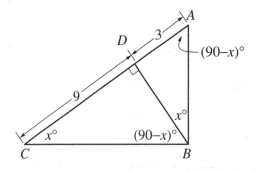

The presence of right triangles suggests that this problem can be approached by using trigonometry. $\overline{BD}$ is in both $\triangle ADB$ and $\triangle BDC$, as is an angle with degree measure x. Thus,

$$\tan x = \frac{3}{BD} \ \text{(in } \triangle ADB)\qquad \text{and} \qquad \tan x = \frac{BD}{9} \ \text{(in } \triangle BDC)$$

Putting the two together, we get $BD^2 = 3 \cdot 9$, or $BD = 3\sqrt{3}$, and $\tan x = \dfrac{\sqrt{3}}{3}$. Therefore, $x = 30$. From $\triangle ABC$ we can see that

$$\sin 30° = \frac{AB}{12} = \frac{1}{2}$$
$$\cos 30° = \frac{BC}{12} = \frac{\sqrt{3}}{2}.$$

Thus, $AB = 6$ and $BC = 6\sqrt{3}$. So $BC + AB + BD = 6\sqrt{3} + 6 + 3\sqrt{3} = 6 + 9\sqrt{3}$ and the correct answer is A.

31. The only procedure in the options that can yield a real root that does not satisfy the original equation is squaring both sides. To see that squaring both sides of an equation can yield an additional root, consider the equation $x - 1 = 2$. This equation has only one root ($x = 3$), but if both sides are squared, the resulting equation is $(x - 1)^2 = 4$, and this equation has two roots, $x = 3$ and $x = -1$.

Therefore, the correct answer is C.

Raising both sides of the equation to the third power could result in an equation that yields <u>complex</u> roots that do not satisfy the original equation, so answer B is incorrect.

32. Since the question asks which <u>must</u> be true, an answer choice can be eliminated by finding a counterexample.

(A) This statement cannot be true, since if n is even, $f(x)$ will have a local minimum at $x = 0$. Consider $f(x) = x^2$.

(B) As seen in (A) above, n cannot be even, and therefore n must be odd.

(C) This statement need not be true, since $f(x) = x^1$ is an example of a function with no local maximum and no local minimum for which n is not divisible by 3.

(D) This statement is not true since n cannot be even.

Therefore, the correct answer is B.

Note: It is not necessary to examine all the options to see that the correct answer is B.

Alternatively, recall or graph the functions $f(x) = x$, $f(x) = x^2$, $f(x) = x^3$, and, $f(x) = x^4$. Of these functions, only $f(x) = x$ and $f(x) = x^3$ has neither a local maximum nor a local minimum. Therefore, A, C, and D need not be true and the correct answer is B.

33. The area of the rectangular region is *hd,* and the area of the semicircular region is $\frac{1}{2}\pi\left(\frac{d}{2}\right)^2$. Since the

area of the semicircular region is $\frac{1}{3}$ of the total area of the window, $\frac{1}{2}\pi\left(\frac{d}{2}\right)^2 = \frac{1}{3}\left[\frac{1}{2}\pi\left(\frac{d}{2}\right)^2 + hd\right]$.

Simplifying this expression yields $\frac{1}{3}\pi\frac{d^2}{4} = \frac{1}{3}hd$ and then solving for *h* gives $\frac{\pi}{4}d = h$.

Therefore, the ratio of *h* to *d* is $\frac{h}{d} = \frac{\pi}{4}$ and the correct answer is C.

34. This equation is equivalent to the matrix equation $\begin{bmatrix} 1 & -1 \\ 1 & 3 \end{bmatrix}\begin{bmatrix} x \\ y \end{bmatrix} = \begin{bmatrix} 2x \\ 2y \end{bmatrix}$. One way to solve this matrix

equation is to write down the equivalent system of linear equations and solve for *x* and *y*. The

equivalent system of equations is $\begin{matrix} x - y = 2x \\ x + 3y = 2y \end{matrix}$. Putting terms involving *x* on one side and terms

involving *y* on the other side, we see that this system is equivalent to the system $\begin{matrix} -x = y \\ x = -y \end{matrix}$.

These two linear equations are clearly equivalent. Therefore, *x* can be any real number *c* as long as

y = −*c*. So the set of solutions to the system of linear equations is the set of vectors $\begin{bmatrix} x \\ y \end{bmatrix}$ of the form

$\begin{bmatrix} x \\ y \end{bmatrix} = \begin{bmatrix} c \\ -c \end{bmatrix}$ where *c* is any real number.

Therefore, the correct answer is C.

Note: $\left\{ \begin{bmatrix} c \\ -c \end{bmatrix}, \text{ where } c \text{ is any real number} \right\}$ is equivalent to the set

$\left\{ \begin{bmatrix} -c \\ c \end{bmatrix}, \text{ where } c \text{ is any real number} \right\}$.

Alternatively, the matrix equation is equivalent to the matrix equation $\begin{bmatrix} 1 & -1 \\ 1 & 3 \end{bmatrix}\begin{bmatrix} x \\ y \end{bmatrix} = \begin{bmatrix} 2 & 0 \\ 0 & 2 \end{bmatrix}\begin{bmatrix} x \\ y \end{bmatrix}$.

Thus $\left(\begin{bmatrix} 1 & -1 \\ 1 & 3 \end{bmatrix} - \begin{bmatrix} 2 & 0 \\ 0 & 2 \end{bmatrix} \right)\begin{bmatrix} x \\ y \end{bmatrix} = \begin{bmatrix} 0 & 0 \\ 0 & 0 \end{bmatrix}$ or $\left(\begin{bmatrix} -1 & -1 \\ 1 & 1 \end{bmatrix} \right)\begin{bmatrix} x \\ y \end{bmatrix} = \begin{bmatrix} 0 & 0 \\ 0 & 0 \end{bmatrix}$.

Multiplying the matrices yields $\begin{bmatrix} -x-y & -x-y \\ x+y & x+y \end{bmatrix} = \begin{bmatrix} 0 & 0 \\ 0 & 0 \end{bmatrix}$.

Since, for the two matrices to be equal, each of the corresponding entries must be equal, it follows

that $x + y = 0$, or $x = -y$, and the solutions are all vectors of the form $\begin{bmatrix} c \\ -c \end{bmatrix}$, where *c* is a real number.

35. The initial amount deposited is $100, so $A(0) = 100$. Every year after the first, the account will gain money from the interest credited plus an additional $50 deposited each year.

So the amount in year n, $A(n)$, will be the amount in year $n - 1$, $A(n - 1)$, <u>plus</u> the interest on the amount in the account in year $n - 1$, $0.08 A(n - 1)$, <u>plus</u> an additional $50 deposit (there is no interest on the $50, since it was just deposited and was not in the account all year).

This situation can be modeled by the recursive equation $A(n) = A(n - 1) + 0.08 A(n - 1) + 50$, which is the same as $A(n) = 1.08 A(n - 1) + 50$.

Therefore, the correct answer is B.

Alternatively, make a table such as the one below, showing the amount in the bank account for the first several years and see which formula, A, B, C or D, gives the same result as the result in the column labeled Amount in Account.

Result of Applying Recursion Formula

n	Amount in Account	A	B	C	D
0	$100	$100	$100	$100	$0
1	$158	$58	$158	$12	$158
2	$220.64	$54.64	$220.64	$4.96	$328.64

Comparing the entries in the table, we can see that the correct answer is B.

36. The derivative of $f(x)$ is $f'(x) = 5x^4 - 21x^2 + 12x - 2$. Graphing this function in the viewing window $[-10, 10] \times [-10, 10]$ produces a graph like the one below.

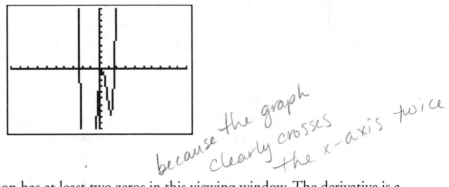

because the graph clearly crosses the x-axis twice

The graph shows that the function has <u>at least two zeros</u> in this viewing window. The derivative is a <u>fourth-degree polynomial</u>, and so can have <u>at most a total of four different zeros</u>.

Note that the graph of the derivative has three turning points, one of which is not in the viewing window but occurs at about $x = -1$ and some value of y less than -10. <u>Since the derivative is a</u>

fourth-degree polynomial, its graph can have at most three turning points. Because of this, we know that all the zeros of the derivative are shown in the viewing window.

If the derivative had another zero outside the viewing window, its graph would have to have another turning point outside the viewing window as well, so that the graph of the derivative could get back to the *x*-axis.

By selecting a smaller viewing window, such as $[-5, 5] \times [-5, 5]$ you can verify that there is no root between 0 and 1, as shown below.

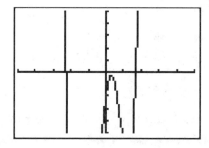

Therefore, $f'(x)$ has two zeros. The correct answer is A.

37. The table tells us that

$$a \times a = b$$
$$a \times b = a$$
$$b \times a = a$$
$$b \times b = a.$$

To answer the question, we need to determine whether the operation $\times$ is commutative or associative or both.

The operation $\times$ is commutative if $x \times y = y \times x$ for all elements x and y in the set. Notice that a and b are the only elements in the set, so, since $a \times a = a \times a$ and $b \times b = b \times b$ we need only check to see whether $a \times b = b \times a$. From the table we can see that $a \times b = a = b \times a$. Thus, the answer will be either C or D.

To determine if the operation $\times$ is associative, we must determine whether $x \times (y \times z) = (x \times y) \times z$ for any choice of x, y, and z in the set. It turns out that the operation defined in the problem is *not* associative, and we can show this by finding a counterexample. To do this, we can choose $x = b$, $y = b$, and $z = a$, then

$$x \times (y \times z) = b \times (b \times a) = b \times a = a, \text{ while}$$
$$(x \times y) \times z = (b \times b) \times a = a \times a = b.$$

This means that for this particular choice of x, y, and z, $x \times (y \times z) \neq (x \times y) \times z$. So the operation $\times$ is *not* associative.

The operation is commutative but not associative; therefore, the correct answer is C.

38. One approach to this problem is to notice that the asymptotes to a hyperbola must pass through the center of the hyperbola. The center of this hyperbola is the point $(x, y) = (1, -2)$. The only line among the options that contains the point $(1, -2)$ is the line $y = x - 3$.

Therefore, the correct answer is A.

An alternative solution is to note that for values of x and y that are large in magnitude, $\dfrac{(x-1)^2}{(y+2)^2} - 1 = \dfrac{1}{(y+2)^2}$. Since the term on the right side of this equation gets close to 0, we can conclude that for values of x and y that are large in magnitude, $\dfrac{(x-1)^2}{(y+2)^2} \approx 1$. This means that $\dfrac{x-1}{y+2} \approx \pm 1$, so the two asymptotes are the lines given by $x - 1 = y + 2$ and $x - 1 = -y - 2$.

The equation $x - 1 = y + 2$ can be rewritten as $y = x - 3$, so the correct answer is A.

39. To determine the radius of the circle, it is useful to put the equation of the circle into standard form: $(x - a)^2 + (y - b)^2 = r^2$, where (a, b) is the center of the circle and r is the radius of the circle.

To put the equation in standard form, it is necessary to complete the square for the terms involving x.

$$x^2 + 2x + y^2 = 0$$
$$(x^2 + 2x + 1) + y^2 = 1$$
$$(x + 1)^2 + y^2 = 1$$

Now that the equation is in standard form, it is evident that the circle has its center at $(-1, 0)$ and has a radius equal to 1.

Therefore, the correct answer is A.

An alternative solution is to graph the upper half of the circle, which corresponds to the function $y = \sqrt{-x^2 - 2x}$ (the upper half of the circle) in an appropriate viewing window of a graphing calculator. From the graph, you can see that the radius of the circle must be 1.

40. In the equation $y = \frac{1}{3}\sin\left(\frac{1}{2}x + \frac{\pi}{3}\right)$, the term $\frac{\pi}{3}$ represents a phase shift of the function $y = \frac{1}{3}\sin\left(\frac{1}{2}x\right)$, so those functions will have the same period. Now the period of $y = \frac{1}{3}\sin\left(\frac{1}{2}x\right)$ can be determined to be 4π by graphing the function in the viewing window $[-4\pi, 4\pi] \times [-1, 1]$.

(This is an appropriate viewing window, since we know the zeros of the sine function are multiples of π and we know that $-\frac{1}{3} \leq y \leq \frac{1}{3}$.)

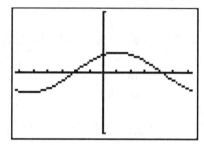

Therefore, the correct answer is D.

Alternatively, recall that the graph of a function like this one will look like a "stretched out" graph of $y = \sin(x)$. Graphing the function $y = \frac{1}{3}\sin\left(\frac{1}{2}x + \frac{\pi}{3}\right)$ in a viewing window $[-2\pi, 2\pi] \times [-4, 4]$ gives the following.

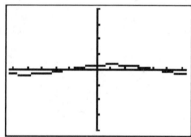

Notice that the graph shows one period of the "stretched out" sine function. Since the range of x-values shown is 4π, it follows that the period of the function is 4π.

Another approach is to use the fact that $y = \frac{1}{3}\sin\left(\frac{1}{2}x\right)$ is a "stretch" of the function $y = \frac{1}{3}\sin(x)$ in the x-direction by a factor of 2. Since the period of $y = \frac{1}{3}\sin(x)$ is 2π, the period of $y = \frac{1}{3}\sin\left(\frac{1}{2}x\right)$ must be $2 \times 2\pi = 4\pi$.

41. I. The range of scores cannot be determined since the lowest score is not known.

II. Although it can be determined that the median must lie in the score group 80–89, we cannot tell exactly what the median score is without knowing more information about the specific scores in that group. For example, if all the scores in that group are 81, the median of all the scores will be 81, but if all the scores in that group are 85, then the median of all the scores will be 85. Therefore, the median score cannot be determined.

III. The mean score cannot be determined without knowing the specific scores, rather than how many are in each group. Therefore, the mean score cannot be determined.

Therefore, neither I nor II nor III can be determined from the information given, and the correct answer is A.

42. Since the inverse function $f^{-1}(x)$ reverses the role of y and x in the equation $y = x^3 + 1$, one method for finding the inverse function is to implement the following two steps.

Step 1: Take the equation $y = f(x)$ and reverse the role of y and x in the equation by replacing each x with y and replacing each y with x.

$$y = f(x)$$
$$y = x^3 + 1$$
$$x = y^3 + 1$$

Step 2: Solve the resulting equation for y to obtain $y = f^{-1}(x)$.

$$x = y^3 + 1$$
$$x - 1 = y^3$$
$$y = \sqrt[3]{x-1}$$
$$f^{-1}(x) = \sqrt[3]{x-1}$$

Since $f^{-1}(x) = \sqrt[3]{x-1}$, the correct answer is B.

Alternatively, recall that the graph of the inverse of a function is the graph of the function "flipped" over the line $y = x$. So, using the graphing calculator, you can graph $f(x) = x^3 + 1$, $y = x$, and each of the options to get the following.

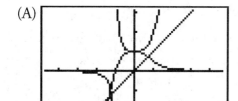

(A)

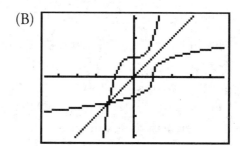

(B)

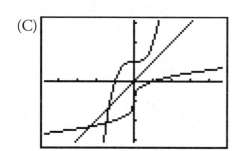

(C)

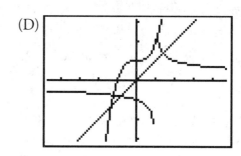

(D)

From the graphs, you can see that the correct answer is B.

43. S is the set of matrices of the form $\begin{bmatrix} a & b \\ c & d \end{bmatrix}$, where each of the entries is a nonzero real number.

For each of the statements I, II, and III, we begin by trying to find an example in which the property is not satisfied. Where possible, constructing examples should be done using numbers that are easy to work with.

  I. Can we find two matrices in S whose product contains at least one entry that is zero: for example, the entry in the first row, first column? Pick as one of the matrices $\begin{bmatrix} 1 & 1 \\ 1 & 1 \end{bmatrix}$, (which is

  easy to work with) and see if you can get a matrix satisfying $\begin{bmatrix} 1 & 1 \\ 1 & 1 \end{bmatrix} \cdot \begin{bmatrix} a & b \\ c & d \end{bmatrix} = \begin{bmatrix} 0 & * \\ * & * \end{bmatrix}$ where $*$

  represents any real number. It follows from the matrix equation that if $a = -c$, the entry in the first row, first column, of the product will be 0. Thus S is not closed under multiplication.

  II. Can we find two matrices A and B in S, where $AB \neq BA$? Set $A = \begin{bmatrix} 1 & 1 \\ 1 & 1 \end{bmatrix}$ and $B = \begin{bmatrix} 1 & 2 \\ 3 & 4 \end{bmatrix}$, which

  has four different entries, so that if the order of multiplying ever affects the product, it is likely to do so with this matrix. Multiply the matrices in both orders and see whether the products are equal.

$$AB = \begin{bmatrix} 1 & 1 \\ 1 & 1 \end{bmatrix} \cdot \begin{bmatrix} 1 & 2 \\ 3 & 4 \end{bmatrix} = \begin{bmatrix} 4 & 6 \\ 4 & 6 \end{bmatrix}$$

$$BA = \begin{bmatrix} 1 & 2 \\ 3 & 4 \end{bmatrix} \cdot \begin{bmatrix} 1 & 1 \\ 1 & 1 \end{bmatrix} = \begin{bmatrix} 3 & 3 \\ 7 & 7 \end{bmatrix}$$

Since the products are not equal, S is not commutative under multiplication.

III. If $\begin{bmatrix} a & b \\ c & d \end{bmatrix}$ is the identity matrix in S, then $\begin{bmatrix} a & b \\ c & d \end{bmatrix} \cdot M = M$ for every matrix M in S. In particular, use the matrices we used in II,

$$\begin{bmatrix} a & b \\ c & d \end{bmatrix} \cdot \begin{bmatrix} 1 & 1 \\ 1 & 1 \end{bmatrix} = \begin{bmatrix} 1 & 1 \\ 1 & 1 \end{bmatrix} \quad \text{and} \quad \begin{bmatrix} a & b \\ c & d \end{bmatrix} \cdot \begin{bmatrix} 1 & 2 \\ 3 & 4 \end{bmatrix} = \begin{bmatrix} 1 & 2 \\ 3 & 4 \end{bmatrix}.$$

Multiplying the left-hand side of both equations gives

$$\begin{bmatrix} a+b & a+b \\ c+d & c+d \end{bmatrix} = \begin{bmatrix} 1 & 1 \\ 1 & 1 \end{bmatrix} \quad \text{and} \quad \begin{bmatrix} a+3b & 2a+4b \\ c+3d & 2c+4d \end{bmatrix} = \begin{bmatrix} 1 & 2 \\ 3 & 4 \end{bmatrix}.$$

In the equations, looking at the first row, first column, gives $a + b = 1$ and $a + 3b = 1$. Therefore, $b = 0$. Since b must be 0, the matrix $\begin{bmatrix} a & b \\ c & d \end{bmatrix}$ is not in S, and therefore S does not contain an identity element.

Based on the answers for I, II, and III above, the correct answer is A.

44. One way to start this problem is to write out a few terms of the sequence:

$$a_1 = 1$$
$$a_2 = a_1 + 2 = 1 + 2$$
$$a_3 = a_2 + 3 = 1 + 2 + 3$$
$$a_4 = a_3 + 4 = 1 + 2 + 3 + 4$$

Continuing in this manner, we can see that $a_n = 1 + 2 + 3 + 4 + \ldots + n$ for every positive integer n.

Since $1 + 2 + 3 + 4 + \ldots + n = \dfrac{n(n+1)}{2}$ for every positive integer n, the correct answer is C.

An alternative solution is to enter the sequence recursively on the graphing calculator and graph it. The resulting graph is clearly not linear, which eliminates B and D as the possible solutions. Examining the formulas for a_n in options A and C, we see that option A gives an incorrect value for a_n (as it does for every other positive integer $n > 1$).

Therefore, the only answer choice that could possibly be the correct answer is C.

45. This function will be undefined only for values of x for which the denominator is equal to 0 or for which the expression inside the square root is negative.

Since $x^2 + 4 > 0$ for every real number x, we need only check whether each of the given values of x results in a 0 in the denominator.

I. If $x = -3$, then $x^3 + x^2 - 5x + 3 = -27 + 9 + 15 + 3 = 0$. Therefore, $f(x)$ is *undefined* when $x = -3$.

II. If $x = -2$, then $x^3 + x^2 - 5x + 3 = -8 + 4 + 10 + 3 = 9$. Therefore, $f(x)$ is *defined* when $x = -2$.

III. If $x = 1$, then $x^3 + x^2 - 5x + 3 = 1 + 1 - 5 + 3 = 0$. Therefore, $f(x)$ is *undefined* when $x = 1$.

Therefore, the correct answer is C.

Alternatively, you can use the calculator to evaluate the function at $x = -3, -2$, and 1. The calculator will give the value of $f(-2)$ and a "Division by zero" error for $f(-3)$ and $f(1)$.

Thus the answer is C.

46.

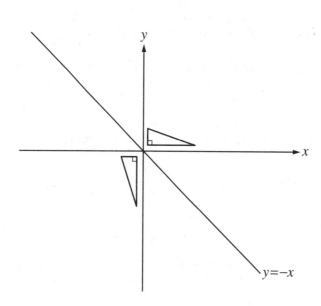

Reflecting the triangle about the line $y = -x$ moves the triangle to the third quadrant with the orientation shown in the figure above.

Rotating the resulting triangle by 90° clockwise about the origin moves the triangle to the second quadrant with the orientation shown in the following figure.

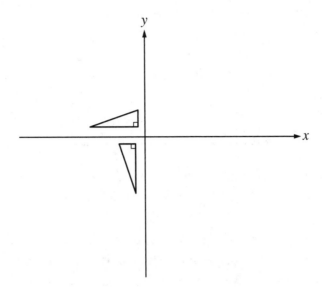

Therefore, the correct answer is D.

Note that first rotating the original triangle by 90° about the origin and then reflecting in the line $y = -x$ will *not* result in the correct answer.

Note also that choices A and B could have been eliminated early on, since the triangles in A and B are not in the second quadrant.

47. The most straightforward way to do this problem is to notice that the quantity we are trying to find, find, $\sum_{i=50}^{75} i^2$, is equal to $\sum_{i=1}^{75} i^2 - \sum_{i=1}^{49} i^2$. This is true since

$$\sum_{i=1}^{75} i^2 = 1^2 + 2^2 + \ldots + (49)^2 + (50)^2 + (51)^2 + \ldots + (75)^2 \qquad \text{and} \qquad \sum_{i=1}^{49} i^2 = 1^2 + 2^2 + \ldots + (49)^2,$$

so subtracting these quantities results in exactly the quantity we are trying to find.

Now use the formula given in the problem.

$$\sum_{i=1}^{75} i^2 - \sum_{i=1}^{49} i^2 = \frac{(75)(76)(151)}{6} - \frac{(49)(50)(99)}{6} = 103{,}025$$

Therefore, $\sum_{i=50}^{75} i^2 = 103{,}025$ and the correct answer is C.

As an alternative solution, a graphing calculator could be used to calculate the sum

$$\sum_{i=50}^{75} i^2 = 103{,}025 \text{ directly.}$$

48.

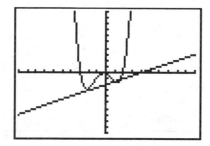

Graphing the fourth-degree polynomial along with the line in the viewing window [–10, 10] × [–10, 10] produces the graph shown above. It is difficult to determine from this graph how many times the line and the quartic polynomial intersect, but we can be sure that they do *not* intersect outside the viewing window. This is true because the graph of a fourth-degree polynomial can have at most three turning points and the graph of this polynomial has all of its turning points inside the viewing window above. In order for the graphs of the two functions to have another point of intersection outside the window, the quartic would have to have another turning point outside the window.

Changing the viewing window to [–4, 2] × [–4, 0] produces the graph below.

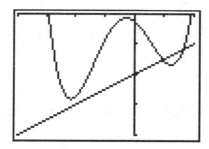

From this figure, it is clear that the graphs of the two functions intersect a total of two times. Note that although the two curves appeared to have another point of intersection in the [–10, 10] × [–10, 10] viewing window, this was due to the poor resolution of the graphing calculator in that viewing window.

Therefore, the correct answer is B.

49.

Number	1	2	3	4	5	6
Probability	$2x$	x	x	x	x	x

In the table above, x is used to represent the probability of throwing a 2. Since the probability of throwing a 2, 3, 4, 5, or 6 is all the same, the probability is x for each. The probability of throwing a 1 is twice the probability of throwing any of the others, so the probability for the number 1 is $2x$. We can determine the value of x by using the fact that the probabilities for all 6 of the numbers must add up to 1.

$$2x + x + x + x + x + x = 1$$
$$7x = 1$$
$$x = \frac{1}{7}$$

In order to determine the probability of throwing a total of 4 in two throws, we must first find all combinations of two throws that add to give 4 and then calculate the probability that each of those combinations will occur. The combinations of throws that result in a sum of 4 are

$$(1, 3) \ (2, 2) \ (3, 1).$$

The probability for each combination is shown below.

$$\text{Probability of } (1, 3) \text{ equals } \left(\frac{2}{7}\right)\left(\frac{1}{7}\right) = \frac{2}{49}.$$

$$\text{Probability of } (2, 2) \text{ equals } \left(\frac{1}{7}\right)\left(\frac{1}{7}\right) = \frac{1}{49}.$$

$$\text{Probability of } (3, 1) \text{ equals } \left(\frac{1}{7}\right)\left(\frac{2}{7}\right) = \frac{2}{49}.$$

Therefore, the total probability of getting a sum of 4 in two throws is $\frac{2}{49} + \frac{1}{49} + \frac{2}{49} = \frac{5}{49}$ and the correct answer is B.

50. Using the definitions of $f(x)$ and $g(x)$, it can be seen that

$$y = f(x) \cdot g(x) = x \cdot g(x) = \begin{cases} -x & \text{if } x < 0 \\ 0 & \text{if } x = 0 \\ x & \text{if } x > 0 \end{cases}.$$

Therefore, $y = |x|$. The only option that is equal to $|x|$ is $f(|x|) = |x|$.

Therefore, the correct answer is D.

Note: Another way to see that $f(x) \cdot g(x) = |x|$ is to graph $f(x) \cdot g(x)$ on the graphing calculator.

Chapter 10

Solutions and Sample Responses for the
Mathematics: Proofs, Models, and Problems, Part 1 Test
and How They Were Scored

▶ ▶ ▶ ▶ ▶ ▶ ▶ ▶ ▶ ▶ ▶ ▶

This chapter presents actual sample responses to the questions in the practice test (chapter 7) and explanations for the scores they received.

As discussed in chapter 5, each question on the *Proofs, Models, and Problems, Part 1* test is scored on a scale from 0 to 5. The general scoring guide used to score these questions is reprinted here for your convenience.

Scoring Guide

Score	Comment

5
- Clearly demonstrates a full understanding of the mathematical content necessary to answer all parts of the question successfully
- Gives a correct and complete response but may contain a minor calculation error

4
- Clearly demonstrates a full understanding of the mathematical content necessary to answer all parts of the question successfully
- EITHER gives a correct and complete response that contains a minor calculation error or misstatement OR gives a correct and almost complete response

3 For a one-part question

- Clearly demonstrates an understanding of all aspects of the question
- Demonstrates the ability to determine an appropriate strategy for answering the question
- Makes substantial progress toward a correct and complete response

For a multipart question

- Clearly demonstrates a full understanding of the mathematical content needed to answer a significant portion of the question successfully
- Gives a correct and complete response to that portion of the question

2 For a one-part question

- EITHER demonstrates a limited understanding of the question OR makes only minimal progress toward a correct and complete response

For a multipart question

- Clearly demonstrates a full understanding of the mathematical content needed to answer a minor portion of the question successfully
- Gives a correct and complete response to that portion of the question

1
- Demonstrates a very limited understanding of the question or questions
- Makes little or no progress toward a correct and complete response

0
- Blank, almost blank, or off topic

Question 1

This section begins with a discussion of the solution to Question 1, followed by several actual responses, with comments from the lead scorers to explain how the scoring guide was used to rate each response.

Question 1–Solution

Both parts of the question ask for the average speed for a 500-mile trip. Average speed for a trip is equal to

$$\frac{\text{total distance traveled}}{\text{total time spent traveling}}.$$

In both parts of the question, the total distance traveled is 500 miles.

(A) The first step in solving this problem is to find the total time Marissa spent traveling, *excluding* the time she spent having lunch. This is equal to the time it took her to travel 250 miles at 55 miles per hour, or $\frac{250}{55}$ hours, plus the time it took her to travel 250 miles at 45 miles per hour, or $\frac{250}{45}$ hours.

Thus her total time spent traveling is $\left(\frac{250}{55} + \frac{250}{45}\right)$ or $\frac{1,000}{99}$ hours.

As discussed above, average speed = $\dfrac{\text{total distance traveled}}{\text{total time spent traveling}}$, so Marissa's average speed for the

500-mile trip is $\dfrac{500}{\frac{1,000}{99}}$ miles per hour. Therefore, the answer is 49.5 miles per hour.

(B) The first step in solving this problem is to find the total time Marissa spent traveling, *including* the time she spent having lunch. This is equal to the time she spent traveling *excluding* the time she spent

having lunch $\left(\frac{1,000}{99}\text{ hours}\right)$, plus the time it took her to have lunch (1 hour), or

$\frac{1,000}{99} + \frac{99}{99} = \frac{1,099}{99}$ hours.

Therefore, Marissa's average speed for the 500-mile trip, including the time she spent having lunch, is

$\dfrac{500}{\frac{1,099}{99}} = \dfrac{500 \cdot 99}{1,099} \approx 45.041$ miles per hour.

Question 1–Sample Responses

Sample Response 1

$Score = 5$

MATHEMATICS: PROOFS, MODELS, AND PROBLEMS, PART 1

Time—60 minutes

4 Questions

Question 1

This question is worth $16\frac{2}{3}$ percent of your score for this test.

Marissa drove 500 miles and stopped only once during the trip. Her average speed for the first 250 miles of the trip was 55 miles per hour. After driving 250 miles, Marissa stopped for 1 hour to have lunch. After her stop Marissa drove at an average speed of 45 miles per hour for the rest of the trip.

(A)   What was Marissa's average speed for the 500-mile trip if the average is calculated <u>excluding</u> the time she spent having lunch? Show how you arrived at your answer.

(B)   What was Marissa's average speed for the 500-mile trip if the average is calculated <u>including</u> the time she spent having lunch? Show how you arrived at your answer.

NOTES

This is the place for scratch work and time to think over the problem before making a fresh start on the next page.

Part A

$\underline{500\ miles}$

$\frac{250}{55}$ $4.\overline{54}$ hours   4 hours 33 min

$.54$ hours = $32.\overline{72}$ minutes

$\frac{250}{45} = 5.\overline{55}$

$.\overline{55}$ min = 33.3 min   5 hours 33 min

Total Time

9 hours 66 min = 10 hours 6 min

Begin your response to question 1 here.

In order to solve the problem remember that D=RT.

Question A:

To determine the time excluding lunch, find the time it took Marissa to travel the first leg of her journey.

$$D=RT \implies \frac{D}{R} = T$$

D = RT

D = 250 miles

R = 55 $\frac{miles}{hour}$

* $\frac{250\,miles}{55\,\frac{miles}{hour}}$ = 4.54 hrs ≈ 4 hours 33 min

* watch units

Then calculate the time traveled for the second part of her journey.

D=RT

* D = 250 miles

R = 45 $\frac{miles}{hour}$

$$D=RT \implies \frac{D}{R} = T$$

$\frac{250\,miles}{45\,\frac{miles}{hour}}$ = 5.5 hrs ≈ 5 hours 33 min

* The rest of the trip was 500-250 = 250 miles.

Her total time, excluding lunch, to travel 500 miles was 10 hrs 6 min.

To find her average speed, divide her total mileage by her total time.

$$D=RT \implies \frac{D}{T} = R$$

D = 500 miles

T = 10 hrs 6 min ≈ 10.1 hrs

$\frac{500\,miles}{10.1\,hrs}$ = 49.5 $\frac{miles}{hr}$ ≈ 50 mph.

Remember, it is your responsibility to demonstrate your competence. This person leaves nothing to chance. The reader knows why each operation is being performed.

(Question 1 continued)

<u>Part B</u>   To determine her average speed including her lunch break, include an extra hour for her total time.

D = 500 miles
T = 11.1 ~~10.1~~ hours

$$D = RT \implies \frac{D}{T} = R$$

$$\frac{500 \text{ miles}}{11.1 \text{ hours}} = 45.\overline{045} \frac{\text{miles}}{\text{hour}}$$

$$\approx 45 \frac{\text{miles}}{\text{hour}}$$

Note the careful attention to units of measure throughout the solution.

Sample Response 2

$$Score = 5$$

MATHEMATICS: PROOFS, MODELS, AND PROBLEMS, PART 1

Time—60 minutes

4 Questions

Question 1

This question is worth $16\frac{2}{3}$ percent of your score for this test.

Marissa drove 500 miles and stopped only once during the trip. Her average speed for the first 250 miles of the trip was 55 miles per hour. After driving 250 miles, Marissa stopped for 1 hour to have lunch. After her stop Marissa drove at an average speed of 45 miles per hour for the rest of the trip.

(A)   What was Marissa's average speed for the 500-mile trip if the average is calculated <u>excluding</u> the time she spent having lunch? Show how you arrived at your answer.

(B)   What was Marissa's average speed for the 500-mile trip if the average is calculated <u>including</u> the time she spent having lunch? Show how you arrived at your answer.

NOTES

$$\frac{250}{55} = \quad x = 4.55 \quad \frac{500}{x+y} = 49.5 \text{ mph.}$$

$$\frac{250}{45} = \quad y = 5.55$$
$$\overline{10.10}$$

$$\frac{500}{x+y+1} =$$

Begin your response to question 1 here.

A) $\dfrac{250 \text{ mi}}{55 \text{ mph}} = 4.55$ hrs section 1

$\dfrac{250}{45} = \dfrac{5.55}{10.1}$ " " 2

Total Time

$\dfrac{500}{10.1} = 49.5$ mph.

B. $\dfrac{500}{10.1 + 1} = \dfrac{500}{11.1} = 45.05$ mph.

Total Time $= 10.1 + 1$

Very terse, but everything is clear through the step-by-step organization.

Sample Response 3

$$Score = 4$$

MATHEMATICS: PROOFS, MODELS, AND PROBLEMS, PART 1

Time—60 minutes

4 Questions

Question 1

This question is worth $16\frac{2}{3}$ percent of your score for this test.

Marissa drove 500 miles and stopped only once during the trip. Her average speed for the first 250 miles of the trip was 55 miles per hour. After driving 250 miles, Marissa stopped for 1 hour to have lunch. After her stop Marissa drove at an average speed of 45 miles per hour for the rest of the trip.

(A) What was Marissa's average speed for the 500-mile trip if the average is calculated <u>excluding</u> the time she spent having lunch? Show how you arrived at your answer.

(B) What was Marissa's average speed for the 500-mile trip if the average is calculated <u>including</u> the time she spent having lunch? Show how you arrived at your answer.

___

NOTES

Haste could well be the culprit here. A minor error with units spoils an otherwise solid effort.

500

$250/55 = 4.5$

$250/45 = 5.6$

$= APP\ 10.1\ HRS. = 10\ HRS\ 6\ MIN$
$+$
$1\ HR\ LUNCH$
___
$11\ HRS\ 6\ MIN.$

6 min is not .6 hr as shown in the denominators below.

EXCLUDING LUNCH $\dfrac{500\ mi}{10.6\ HR} = 49.5\ MPH.$

However, here 10.1 was actually used to divide,

INCLUDING LUNCH $\dfrac{500\ m}{11.6\ HRS} = 43.1\ MPH$

but here dividing by 11.6 gives an incorrect answer.

Begin your response to question 1 here.

Nothing much here except restating the answers.

MARISSA'S AVERAGE SPEED FOR THE 500 MILE TRIP EXCLUDING THE TIME SHE SPENT HAVING LUNCH WOULD BE APPROXIMATELY 49.5 MPH. I ARRIVED AT THIS ANSWER BY KNOWING RT. = DISTANCE. FIRST I FOUND THE AVERAGE SPEED FOR THE FIRST 250 MILES. $250 \frac{mi}{55 \, mph} = 4.5$.

NEXT WAS THE FOLLOWING 250 MILES. $250 \frac{mi}{45 \, mph} = 5.6$.

B. HER AVERAGE SPEED FOR INCLUDING LUNCH WAS 43.1 MPH

The error might have been caught if work in the notes area was rewritten in a better form with some narative.

Sample Response 4

$$Score = 4$$

MATHEMATICS: PROOFS, MODELS, AND PROBLEMS, PART 1

Time—60 minutes

4 Questions

Question 1

This question is worth $16\frac{2}{3}$ percent of your score for this test.

Marissa drove 500 miles and stopped only once during the trip. Her average speed for the first 250 miles of the trip was 55 miles per hour. After driving 250 miles, Marissa stopped for 1 hour to have lunch. After her stop Marissa drove at an average speed of 45 miles per hour for the rest of the trip.

(A)   What was Marissa's average speed for the 500-mile trip if the average is calculated excluding the time she spent having lunch? Show how you arrived at your answer.

(B)   What was Marissa's average speed for the 500-mile trip if the average is calculated including the time she spent having lunch? Show how you arrived at your answer.

NOTES

500.

250 / 55

500 250 = 250 / 45

Begin your response to question 1 here.

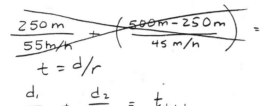

$$t = d/r$$

(A) $\dfrac{d_1}{r_1} + \dfrac{d_2}{r_2} = t_{total}$

$d_1 = 500\,m$ $d_2 = $ total distance $-$ distance traveled $= 500\,m - 250$

$r_1 = 55\,m/h$ $r_2 = 45\,m/h$

Clearly and carefully finds time
but doesnot proceed and find the
rate as well.

$\dfrac{250\,m}{55\,m/h} + \dfrac{250}{45\,m/h} =$

$4.55\,h + 5.56\,h = 10.1\,hours$

But does so in Part B so the
omission is considered minor.

(B) Average Speed $= \dfrac{\text{total distance}}{\text{total time}}$ (driving time + lunch stop)

$= \dfrac{500\,m}{11.1\,h}$

$= 45\,m/h$

Always reread the question to
make certain you answer the
question that was asked and
showed everything that was required.

Sample Response 5

$$\text{Score} = 3$$

MATHEMATICS: PROOFS, MODELS, AND PROBLEMS, PART 1

Time—60 minutes

4 Questions

<u>Question 1</u>

This question is worth $16\frac{2}{3}$ percent of your score for this test.

Marissa drove 500 miles and stopped only once during the trip. Her average speed for the first 250 miles of the trip was 55 miles per hour. After driving 250 miles, Marissa stopped for 1 hour to have lunch. After her stop Marissa drove at an average speed of 45 miles per hour for the rest of the trip.

(A)   What was Marissa's average speed for the 500-mile trip if the average is calculated <u>excluding</u> the time she spent having lunch? Show how you arrived at your answer.

(B)   What was Marissa's average speed for the 500-mile trip if the average is calculated <u>including</u> the time she spent having lunch? Show how you arrived at your answer.

---

d_T is the total distance traveles
d_a = is distance for first part of trip.
d_b = is distance for second part of trip
r_a = rate for first part of trip.
r_b = rate for second part of trip.

NOTES

(d_r)
distance

A graph is often a good start to visualizing the problem and beginning to formulate a strategy to solve it.

Given:
$d_T = 500\,mi$

$d_A = 250\,mi \quad d_1 = 250\,mi$
$Ar_2 = 55\,mi/hr \quad Ar_2 = 45\,mi/hr$

Excluding the hour for lunch the average speed would be:

$$\frac{55\,mi/h + 45\,mi/hr}{2}$$

However, Part A is found incorrectly by averaging rates—rather than using the quantities "distance" and "time" that label the axes of the graph.

Begin your response to question 1 here.

d_T = total distance of trip = 500 mi.

d_1 = distance for first part of trip = 250mi

d_2 = distance for second part of trip = 250 mi.

Ar_1 = average rate for first part of trip = 55 mph

Ar_2 = average rate for second part of trip = 45 mph

Ar_T = total average rate for trip

A) since we are not including the 1 hour lost during lunch we find:

$$\therefore \quad Ar_T = \frac{Ar_1 + Ar_2}{2} = \frac{55\,\text{mph} + 45\,\text{mph}}{2} = 50\,\text{mph}$$

Part B is begun afresh and solved completely and correctly.

B) since we need a total time value to calculate the trip time we need first to determine the time needed to complete trip 1.

trip 1 time = T_1 trip 2 time = T_2 T_T = total trip time

$$T_1 = \frac{250\,\text{mi}}{1} \times \frac{1\,\text{hour}}{55\,\text{mi}} = 4.55 \text{ hours}$$

$$T_2 = \frac{250\,\text{mi}}{1} \times \frac{1\,\text{hour}}{45\,\text{mi}} = 5.55 \text{ hours}$$

$$T_0 = 1 \text{ hour} \quad (\text{Given})$$

$T_T = T_1 + T_2 + T_0$

$T_T = 4.55\,\text{hr} + 5.55\,\text{hr} + 1\,\text{hr}$

$T_T = 11.1 \text{ hr}$

$$\therefore \quad \text{Average speed} = \frac{d_T}{T_T} = \frac{500\,\text{mi}}{11.1\,\text{hr}} = 45 \text{ miles per hour.}$$

Sample Response 6

$$Score = 2$$

MATHEMATICS: PROOFS, MODELS, AND PROBLEMS, PART 1

Time—60 minutes

4 Questions

Question 1

This question is worth $16\frac{2}{3}$ percent of your score for this test.

Marissa drove 500 miles and stopped only once during the trip. Her average speed for the first 250 miles of the trip was 55 miles per hour. After driving 250 miles, Marissa stopped for 1 hour to have lunch. After her stop Marissa drove at an average speed of 45 miles per hour for the rest of the trip.

(A)   What was Marissa's average speed for the 500-mile trip if the average is calculated <u>excluding</u> the time she spent having lunch? Show how you arrived at your answer.

(B)   What was Marissa's average speed for the 500-mile trip if the average is calculated <u>including</u> the time she spent having lunch? Show how you arrived at your answer.

NOTES

$D = 500$

500   AVG S   55
          FOR 250

AVC SPEED
AS $L = 1\,hr$

                  250   45

$D = 250$ 250

$AS = .55$ $AS = 45$

$\frac{1}{2}$ TIME @ 55   50 mph

$\frac{1}{2}$ TIM. @ $\frac{45}{100}$

Begin your response to question 1 here.

(TOTAL DISTANCE)

TD = 500

A)

AVERAGE SPEED 1ST 250 miles

AS 1ST = 55 mph

AVERAGE SPEED 2 = 250 miles

AS 2ND = 45 mph

X = TOTAL AVERAGE SPEED

$$\frac{55 + 45}{2} = X$$

Averaging rates gives
the wrong average rate.

50 = X

50 mph = AVERAGE SPEED

Presumedly the 10 hours is found using the rate
in Part A, but Part B is done correctly with that time.

B) IT TOOK 10 hours to drive
the total distance. IF include
lunch, then IT TOOK 11 hours

$$11 \overline{\smash{\big)}\ 500} \quad \frac{45.45}{}$$
44
60
55
50
44
60

AVERAGE SPEED WAS

45.45 mph

when lunch break
was included.

Sample Response 7

$$\text{Score} = 2$$

MATHEMATICS: PROOFS, MODELS, AND PROBLEMS, PART 1

Time—60 minutes

4 Questions

<u>Question 1</u>

This question is worth $16\frac{2}{3}$ percent of your score for this test.

Marissa drove 500 miles and stopped only once during the trip. Her average speed for the first 250 miles of the trip was 55 miles per hour. After driving 250 miles, Marissa stopped for 1 hour to have lunch. After her stop Marissa drove at an average speed of 45 miles per hour for the rest of the trip.

(A)   What was Marissa's average speed for the 500-mile trip if the average is calculated <u>excluding</u> the time she spent having lunch? Show how you arrived at your answer.

(B)   What was Marissa's average speed for the 500-mile trip if the average is calculated <u>including</u> the time she spent having lunch? Show how you arrived at your answer.

NOTES

$$\frac{55m}{h} \qquad \frac{250}{55} = 4.5 \text{ hours}$$

$$\frac{250}{45} = 5.6 \text{ hours}$$

$$r(\qquad = 500$$

$$r_1 t_1 = r_2 t_2$$

$$55 t_1 = \cancel{5} \, 45 t_2$$

$$\frac{t_1}{t_2} = 0.\overline{81}$$

Begin your response to question 1 here.

(A) excluding lunch :   The times are found correctly, but

@ 55 mph drove for 4.5 hrs $\left(\frac{250}{55}\right)$

@ 45 mph drove for 5.6 hrs $\left(\frac{250}{45}\right)$

there is no evidence that the times
have been used to find the average.

50 mph average

(B) including lunch

4.5 hrs + 5.6 hrs + 1 hr. = 11.1 hrs

once again the correct time is
found, but no attempt is made
to find the rate.

Sample Response 8

$Score = 1$

MATHEMATICS: PROOFS, MODELS, AND PROBLEMS, PART 1

Time—60 minutes

4 Questions

Question 1

This question is worth $16\frac{2}{3}$ percent of your score for this test.

Marissa drove 500 miles and stopped only once during the trip. Her average speed for the first 250 miles of the trip was 55 miles per hour. After driving 250 miles, Marissa stopped for 1 hour to have lunch. After her stop Marissa drove at an average speed of 45 miles per hour for the rest of the trip.

(A)   What was Marissa's average speed for the 500-mile trip if the average is calculated <u>excluding</u> the time she spent having lunch? Show how you arrived at your answer.

(B)   What was Marissa's average speed for the 500-mile trip if the average is calculated <u>including</u> the time she spent having lunch? Show how you arrived at your answer.

NOTES

A) Avg. speed for 1st part = 55 mph
   Avg. speed for 2nd part = 45 mph
   Avg. speed for entire trip = 50 mph

B) include lunch

Begin your response to question 1 here.

A) The total trip was 500 miles. Marissa drove an average speed of 55 mph during the 1st part of the trip. She drove an average speed of 45 mph during the 2nd part of the trip.

Therefore she drove an average speed of $\frac{45 + 55}{2} =$ (50 mph) during the 500 mile trip. Averaging rates!

B) The fact that she had lunch and stopped for 1 hour would affect her average speed in the following way:

Avg. speed for 1st part = 55 mph
Avg. speed during lunch = 0 mph
Avg. speed during 2nd part = 45 mph

If you include lunch, then you have to total all of the average speeds (55 + 0 + 45 = 100) and divide by the 3 segments of the trip (1st part, lunch, and 2nd part) Thus, 100 ÷ 3 = 33.33 mph. Thus, the average speed for the entire trip is

(33.33 mph)

Averaging rates again, this time including the zero rate during lunch break.

No evidence that distance and time are essential in this problem.

Question 2

This section begins with a discussion of the solution to Question 2, followed by several actual responses, with comments from the lead scorers to explain how the scoring guide above was used to rate each response.

Question 2–Solution

(A) To move the graph of $y = f(x)$ up 4 units, replace the variable y with $(y - 4)$ so that the desired function is $y - 4 = x^2$, or $y = x^2 + 4$.

If, while solving this problem, you are unsure whether the correct answer is $y = x^2 + 4$, or $y = x^2 - 4$, you could graph both on the graphing calculator and observe which function produces the desired graph.

Solutions for part (B) can be tested similarly using the graphing calculator.

(B) To move the graph of $y = f(x)$ to the right 3 units, replace the variable x with $(x - 3)$, so that the desired function is $y = (x - 3)^2$, or $y = x^2 - 6x + 9$.

(C) Doubling the x-coordinate of each point of the graph is the same as "stretching" the graph horizontally by a factor of 2. To do this, replace the variable x with $\left(\frac{1}{2}x\right)$.

Thus, the desired equation is or $y = \left(\frac{1}{2}x\right)^2$, or $y = \frac{1}{4}x^2$.

While solving this problem, you may be unsure whether the $y = (2x)^2$, or $y = \left(\frac{1}{2}x\right)^2$ will produce the correct graph. Once again, both functions can be graphed on the calculator, and the graphs can be compared to the graph of $y = x^2$ to see which one produces the desired result.

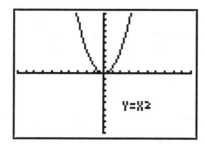

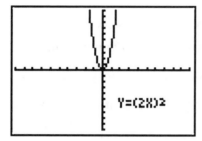

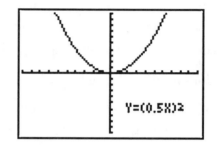

The points with the same y-coordinate as $y = x^2$ have <u>one-half</u> the x-coordinate, so $y = (2x)^2$ does *not* produce the desired graph.

This graph has the desired effect.

Therefore, the correct answer is $y = \left(\dfrac{1}{2}x\right)^2$ or $y = \dfrac{1}{4}x^2$.

Question 2–Sample Responses $Score = 5$

Sample Response 9

This question is worth $16\frac{2}{3}$ percent of your score for this test.

(A)   Determine an equation of the function whose graph is formed by moving each point on the graph of the function $y = x^2$ up 4 units. Show how you arrived at your answer.

(B)   Determine an equation of the function whose graph is formed by moving each point on the graph of the function $y = x^2$ to the right 3 units. Show how you arrived at your answer.

(C)   Determine an equation of the function whose graph is formed by moving each point on the graph of the function $y = x^2$ to a point with the same y-coordinate and twice the x-coordinate. Show how you arrived at your answer.

NOTES

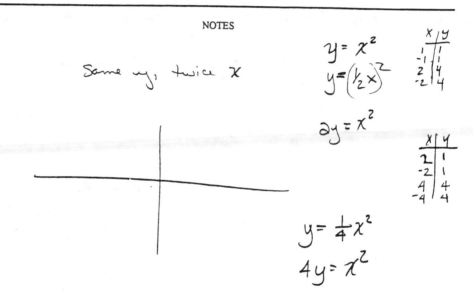

Same y, twice x

$y = x^2$

$y = \left(\tfrac{1}{2}x\right)^2$

X	Y
1	1
-1	1
2	4
-2	4

$2y = x^2$

X	Y
2	1
-2	1
4	4
-4	4

$y = \tfrac{1}{4}x^2$

$4y = x^2$

Begin your response to question 2 here.

Begins with a clear general statement and an algebraic form of an appropriate parabola.

The general form of a parabola is

$$(y-k) = \frac{1}{4p}(x-h)^2 \text{ where the}$$

parabola is vertical. $\left(\frac{\cup}{4p \text{ pos}} \text{ or } \frac{\cap}{4p \text{ neg}}\right)$.

The h and k values determine movement along x & y respectively of the vertex.

The graph of $y = x^2$ has a vertex at $(0,0)$

$p = \frac{1}{4}$

Shows that the reference curve is of this form.

Ⓐ moving the graph 4 units up means the vertex of the graph is at $(0,4)$. Therefore, the graph

$\underset{h,k}{}$

would look like this:

The equation would be:

$$(y-4) = x^2$$

Considers the effect of the

specific movement described on the general form of the equation.

(Question 2 continued)

Similarly, here another translation.

Ⓑ Moving the graph 3 units to the right, means the vertex of the graph would be (3,0). The graph:
$\underset{h,k}{(3,0)}$

The equation would be:
$y = (x-3)^2$

In part C, the x-variable is replaced by ½x, then a table, a graph, and a word description are used to convince ~~An equation~~ where the y coordinate remained the reader of the writer's understanding.

Ⓒ ~~An equation~~ where the y coordinate remained the same, but the x coordinate was twice its' original value would deal with the coefficient of x^2.

~~From~~ $y = x^2$ $(y = 1x^2)$, we must find a coefficient of x^2 to satisfy the condition. $y = (\tfrac{1}{2}x)^2,$ $\boxed{y = \tfrac{1}{4}x^2}$ ← p = 1

or

$4y = x^2$

as p increases, parabola gets wider

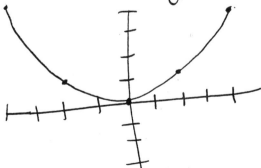

x	y
2	1
-2	1
4	4
-4	4	← origin still
0	0	← vertex

Sample Response 10

$$\text{Score} = 5$$

Question 2

This question is worth $16\frac{2}{3}$ percent of your score for this test.

(A) Determine an equation of the function whose graph is formed by moving each point on the graph of the function $y = x^2$ up 4 units. Show how you arrived at your answer.

(B) Determine an equation of the function whose graph is formed by moving each point on the graph of the function $y = x^2$ to the right 3 units. Show how you arrived at your answer.

(C) Determine an equation of the function whose graph is formed by moving each point on the graph of the function $y = x^2$ to a point with the same y-coordinate and twice the x-coordinate. Show how you arrived at your answer.

NOTES

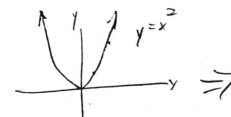
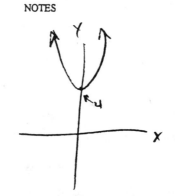

In this approach, a narrative describes what is happening to the y-values, and an equation is produced with this relationship.

Begin your response to question 2 here.

A) $y = x^2$

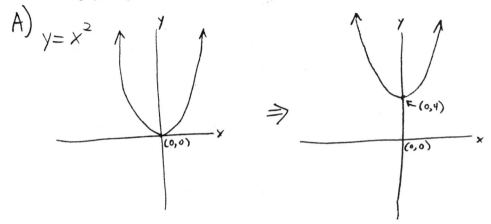

To find the new equation, we must determine what has taken place. For every value of x, y has increased by the constant value of 4 over the original value of x^2. Therefore, the new equation is $Y = x^2 + 4$

B) $y = x^2$

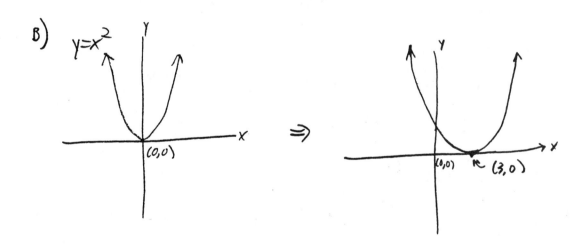

(Question 2 continued)

To find the new equation we must determine what has taken place. For every value of x, y now has the value of x^2 using the x that is three units less. Therefore, $y = (x-3)^2$

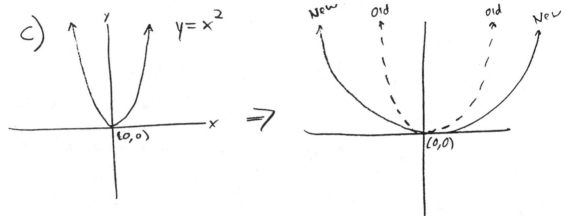

c) $y = x^2$ ... $(0,0)$ ... $\Rightarrow$ ... New ... Old ... Old ... New ... $(0,0)$

To find the new equation, we must determine what has taken place. For every value of y, x is equal to twice the previous value. The previous value of x was $x = \sqrt{y}$. Now $x = 2\sqrt{y}$ therefore, the new equation is $y = \dfrac{x^2}{4}$

Notice, in the last part, one must consider what is happening to the x-values.

Sample Response 11

Score = 5

<u>Question 2</u>

This question is worth $16\frac{2}{3}$ percent of your score for this test.

(A) Determine an equation of the function whose graph is formed by moving each point on the graph of the function $y = x^2$ up 4 units. Show how you arrived at your answer.

(B) Determine an equation of the function whose graph is formed by moving each point on the graph of the function $y = x^2$ to the right 3 units. Show how you arrived at your answer.

(C) Determine an equation of the function whose graph is formed by moving each point on the graph of the function $y = x^2$ to a point with the same y-coordinate and twice the x-coordinate. Show how you arrived at your answer.

NOTES

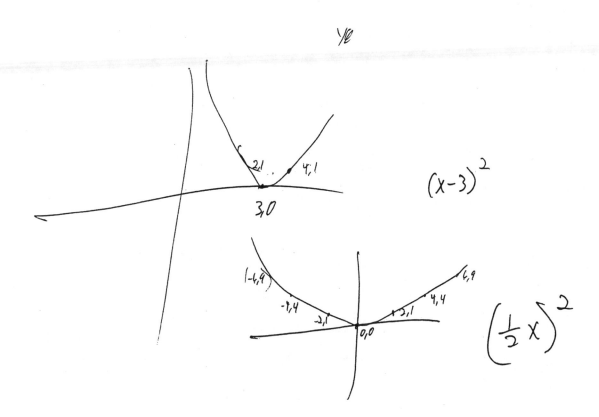

A minimalist version because the solution must explain how the equation is obtained.

Begin your response to question 2 here.

A) Raising a graph up 4 units means increasing the y value by 4

$$\therefore \quad if \quad y = x^2$$

Then Raising by four $\Rightarrow$ $\underline{\underline{y = x^2 + 4}}$

B)

In order to $(x-3)$

shift X but maintain y

you must compensate for the changed X before the square

$(x-3)$ compensates for the shift in X

$$\underline{y = (x-3)^2}$$
Positive 3

$y = 9 - 9$

C)

$(\frac{1}{2} x)$ compensates for a doubled X value

$$\underline{y = (\frac{1}{2}x)^2}$$

The graphs, while probably not necessary, do reinforce that the terse explanations are evidence of a clear understanding of the concepts in this problem.

Graphs

A)

B)

C)

Sample Response 12

Score = 4

Begin your response to question 2 here.

A) I need to find the equation for the function, f(x), moving up 4 pts.

First, I will graph $f(x) = x^2$.

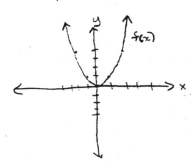

By building a table:

x	f(x)
1	1
2	4
0	0
-1	1
-2	4

Now, if I move each point up 4 units, the graph will shift up 4 units.

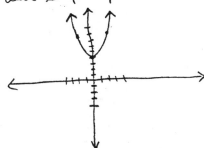

And by building another table:

x	f(x) + 4
1	1 + 4 = 5
2	4 + 4 = 8
0	0 + 4 = 4
-1	1 + 4 = 5
-2	4 + 4 = 8

Therefore, the equation for this graph is: $f(x) = x^2 + 4$

Both part A and part B are explained in the table, where the indicated operations produce the required graph.

(Question 2 continued)

3) Next, $\overset{(I\ will)}{\wedge}$ shift the graph to the right 3 units,

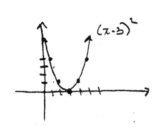

Building a table

x	$f(x)$
3	0
4	1
2	1
1	4
5	4

Using the information from the table and the graph, the equation for this function will be:

$$f(x) = (x-3)^2$$

By testing the pts in the table, I can verify the equation.

$$f(1) = (1-3)^2 = 4$$
$$f(4) = (4-3)^2 = 1$$
$$f(2) = (2-3)^2 = 1$$
$$f(5) = (5-3)^2 = 4$$

Hence the equation will be,

$$f(x) = (x-3)^2$$

(Question 2 continued) But in part c , the table fails to indicate a connection between x and f(x) by indicated operations or otherwise.

c) Now, I will find the eq. for moving each pt, to a new pt, with the same y-coordinate and twice the x coordinate.

By graphing $f(x) = x^2$, I have :

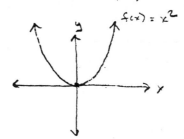

The explanation of this last part was found wanting. It is not all that clear how the resulting equation was found from the table given. It is the responsibility of the examinee to demonstrate or her understanding by explaining the process of " I see that."

First, I will make a table where :

x^2	x	$2x$
2	4	8
1	1	2
0	0	0
-1	1	2
-2	4	8

By evaluating the table, I see that :

$$f(x) = \left(\frac{x}{2}\right)^2$$

By graphing,

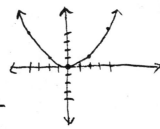

Hence, $f(x) = \left(\frac{x}{2}\right)^2$

Sample Response 13

$Score = 3$

Question 2

This question is worth $16\frac{2}{3}$ percent of your score for this test.

(A)   Determine an equation of the function whose graph is formed by moving each point on the graph of the function $y = x^2$ up 4 units. Show how you arrived at your answer.

(B)   Determine an equation of the function whose graph is formed by moving each point on the graph of the function $y = x^2$ to the right 3 units. Show how you arrived at your answer.

(C)   Determine an equation of the function whose graph is formed by moving each point on the graph of the function $y = x^2$ to a point with the same y-coordinate and twice the x-coordinate. Show how you arrived at your answer.

NOTES

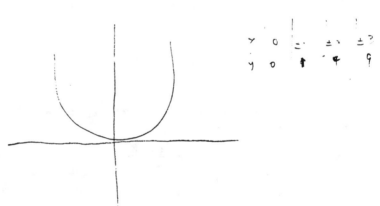

$2(y) = x^2$

$2y = x^2$

$y = \frac{1}{2}x^2$

$y = x^2$

$y = \frac{1}{2}x^2$

Parts   A and B are satisfactory.

Begin your response to question 2 here.

A) By moving the function $y = x^2$ up by 4 units

$$f(x) = x^2$$

$$g(x) = f(x) + 4$$

$$\therefore \quad y = x^2 + 4$$

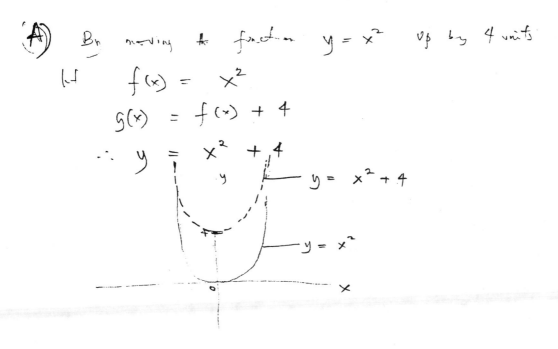

$y = x^2 + 4$

$y = x^2$

Equation of the function $= y = x^2 + 4$

b) $f(x) = x^2$ left to new function by $g(x)$

$$g(x) = f(x-3) = (x-3)^2$$

$$g(x) = (x-3)^2$$

Equation of the new function $= \quad y = (x-3)^2$

$y = (x-3)^2$

But in Part C, the new function is $\frac{1}{2}$ of the original
so the same x will produce $\frac{1}{2}$ of y instead
of the same y paired with twice the x-value.

Examinee clearly misunderstands a part of this problem.

(Question 2 continued)

(c) $f(x) = x^2$

$4 \quad g(x) = \frac{1}{2} f(x) = \frac{1}{2} x^2$

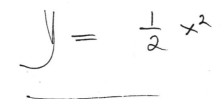

$g(x) = \frac{1}{2} x^2$

$y = \frac{1}{2} x^2$

The new equation will be

$$y = \frac{1}{2} x^2$$

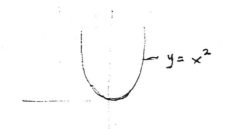

$y = x^2$

Sample Response 14

$Score = 3$

<u>Question 2</u>

This question is worth $16\frac{2}{3}$ percent of your score for this test.

(A)   Determine an equation of the function whose graph is formed by moving each point on the graph of the function $y = x^2$ up 4 units. Show how you arrived at your answer.

(B)   Determine an equation of the function whose graph is formed by moving each point on the graph of the function $y = x^2$ to the right 3 units. Show how you arrived at your answer.

(C)   Determine an equation of the function whose graph is formed by moving each point on the graph of the function $y = x^2$ to a point with the same y-coordinate and twice the x-coordinate. Show how you arrived at your answer.

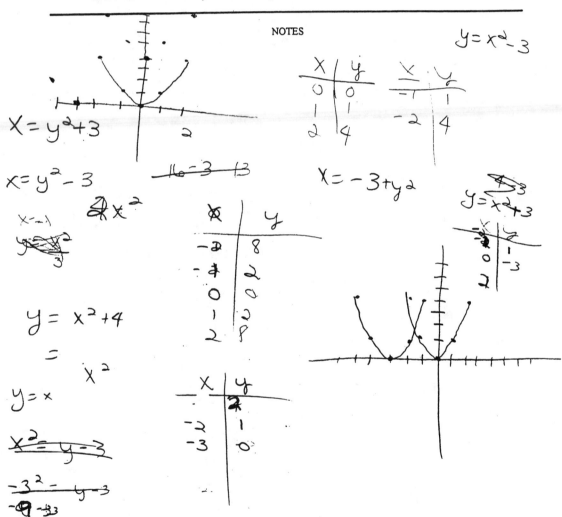

Part A and Part C are satisfactory.

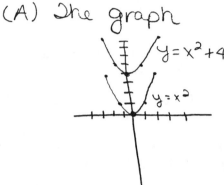

Begin your response to question 2 here.

(A) The graph

$y = x^2 + 4$

$y = x^2$

$y = x^2$

x	y
-2	4
-1	1
0	0
1	1
2	4

$y = x^2 + 4$

x	y
-2	8
-1	5
0	4
1	5
2	8

By looking at the graph, I try to figure an equation that the origin is at (0,4). Then by using an x-y table I plotted points which led me to the equation $y = x^2 + 4$

(B)   The graph

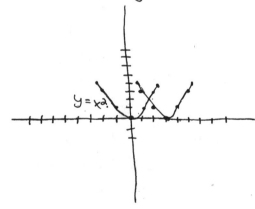

$y = x^2$

I want an equation where the origin is at (3,0).

Using an x-y table

y	x
7	2
4	1
3	0
4	-1
7	-2

$y = x^2 + 3$

Notice here that equation would shift 3 units in the y-direction, not the x-direction. Further, if the table of values had been graphed correctly this would have been very clear.

(Question 2 continued)

(C)

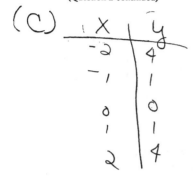

X	y
-2	4
-1	1
0	0
1	1
2	4

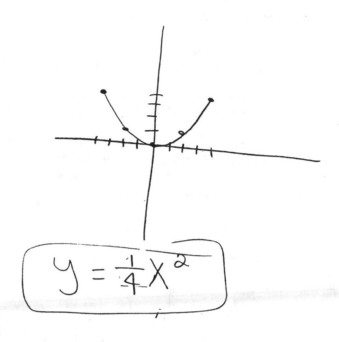

2x	y
-4	4
-2	1
0	0
2	1
4	4

$$y = \frac{1}{4}X^2$$

I figured it out by looking at the graph along with an x-y coordinate table.

Sample Response 15

Score = 2

<u>Question 2</u>

This question is worth $16\frac{2}{3}$ percent of your score for this test.

(A)   Determine an equation of the function whose graph is formed by moving each point on the graph of the function $y = x^2$ up 4 units. Show how you arrived at your answer.

(B)   Determine an equation of the function whose graph is formed by moving each point on the graph of the function $y = x^2$ to the right 3 units. Show how you arrived at your answer.

(C)   Determine an equation of the function whose graph is formed by moving each point on the graph of the function $y = x^2$ to a point with the same y-coordinate and twice the x-coordinate. Show how you arrived at your answer.

NOTES

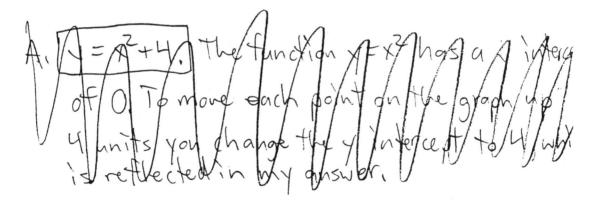

A. $y = x^2 + 4$. The function $y = x^2$ has a y-interc of 0. To move each point on the graph up 4 units you change the y intercept to 4 which is reflected in my answer.

B.

Part A is satisfactory.

Begin your response to question 2 here.

A. $\boxed{y = x^2 + 4}$

$y = x^2$ has a y intercept of 0. To move each point up 4 units you need to change the y intercept to 4, which is reflected in my answer.

Neither Part B nor Part C is correct,

B. $\boxed{y = x^2 + 9}$, when the x intercept = 3, y = 9, $3^2 = 9$.

and the explanations exhibit serious conceptual flaws.

C. $\boxed{y = 2x^2}$ the y intercept stays at 0, however in order to change the graph to twice the x coordinate, you need to multiply the x^2 by 2.

Notice in Part C that the stated answer would double the y-value not the x-value.

Question 3

This section begins with a discussion of the solution to Question 3, followed by several actual responses, with comments from the lead scorers to explain how the scoring guide above was used to rate each response.

Question 3–Solution

Since City A's population is projected to increase at a rate of 3% per year, its population x years from now is projected to be $250,000(1.03)^x$. City A's population x years from now can be modeled by the function $y = 250,000(1.03)^x$.

Since City B's population is projected to increase at 1% per year, its population x years from now is projected to be $350,000(1.01)^x$. City B's population x years from now can be modeled by the function $y = 350,000(1.01)^x$.

From this point, the solution can continue several ways.

Method 1
The two populations will be equal when $250,000(1.03)^x = 350,000(1.01)^x$.

It follows that:

$$\frac{250,000}{350,000} = \frac{1.01^x}{1.03^x}$$

$$\frac{5}{7} = \left(\frac{1.01}{1.03}\right)^x$$

$$\log\left(\frac{5}{7}\right) = x\log\left(\frac{1.01}{1.03}\right)$$

$$x \approx 17.15$$

Thus the population will be equal in approximately 17 years. At that time both populations will be approximately $250,000(1.03)^{17}$ or $415,165$.

Method 2
The populations are equal for the value of x at which the graphs of the functions $y = 250,000(1.03)^x$ and $y = 350,000(1.01)^x$ intersect.

Since at $x = 0$ the value of the functions are 250,000 and 350,000 and, at the rates of increase given, the populations will probably take years to become equal, a viewing window on the graphing calculator needs to be specified to encompass a reasonable number of years and population size. If you assume that the population will be equal within 30 years and the population size at that time will be under

1,000,000, then the viewing window you would select would be $[0, 30] \times [0, 1,000,000]$. Using the graphing calculator to graph the functions using this viewing window gives the following graph:

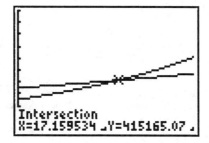

Using the calculator to calculate the point of intersection of the two graphs yields (17.15934, 415,165). Therefore, the population of the two cities will be equal approximately 17.15 years from now. The population at that time will be approximately 415,165.

If you mistakenly assume that populations will be equal within 10 years, the population size at that time would be under 400,000, and your viewing window would be $[0, 10] \times [0, 400,000]$. Using the graphing calculator to graph the functions using this viewing window gives the following graph:

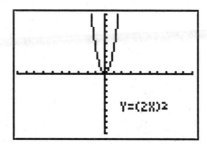

This viewing window does not include the point of intersection of the two graphs, so you would need to try to expand the viewing range. If you tried, say, $[0, 20] \times [0, 500,000]$, you would get the following graph:

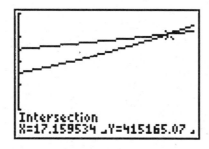

Using the calculator to calculate the point of intersection of the two graphs yields (17.15934, 415,165). Therefore, the population of the two cities will be equal approximately 17.15 years from now. The population at that time will be approximately 415,165.

Question 3–Sample Responses

Score=5

Sample Response 16

Question 3

This question is worth $33\frac{1}{3}$ percent of your score for this test.

City A currently has a population of 250,000 and City B currently has a population of 350,000. If the population of City A increases at a constant rate of 3% per year and the population of City B increases at a constant rate of 1% per year, then in approximately how many years will the population of the two cities be equal? According to this projection, what will the population of City A be at that time? Explain how you arrived at your answer.

NOTES

$$\text{Population } A = (250,000)(1.03)^y$$
$$B = (350,000)(1.01)^y$$

$$2.97035$$
$$.017 = .06$$
$$3375$$

Begin your response to question 3 here.

The population of each city on a given year is:

Population clearly set up as exponential growth.

$P_A = $ *population of City A* $= (250,000)(1.03)^y$ [1]

$P_B = $ *Population of City B* $= (350,000)(1.01)^y$ [2]

where $y=0$ *is the current year*
      $y=1$ *is next year, etc...*

The cities will have the same population when $P_A = P_B$

     —or—

The populations are set equal and a solution for the time (y) is sought.

$P_A = (250,000)(1.03)^y = (350,000)(1.01)^y = P_B$ [3]

Solving for Y, we got

$$(1.03)^y = \frac{(350,000)(1.01)^y}{250,000}$$

Clearly set up, and solved, and answered.

$$(1.03)^y = \left(\frac{350,000}{250,000}\right)(1.01)^y$$

$$\frac{(1.03)^y}{(1.01)^y} = \left(\frac{350,000}{250,000}\right) \implies 1.4$$

$$\left(\frac{1.03}{1.01}\right)^y = 1.4$$

$$(1.0198)^y \cong 1.4$$

$$\ln\left[(1.0198)^y\right] \cong \ln(1.4)$$

$$y\left[\ln(1.0198)\right] \cong \ln(1.4)$$

After finding the time, the population at that time is calculated.

$$y = \frac{\ln(1.4)}{\ln(1.0198)} = \frac{(0.3364+7)}{(0.019606)} = 17 \text{ years}$$

At that time, the population of city A will be...

$$P_A = (250,000)(1.03)^{17} = 413,212.$$

If the solution had stopped with 17 years, the paper would have earned a 4 instead of a 5 score.

Sample Response 17

$Score = 5$

<u>Question 3</u>

This question is worth $33\frac{1}{3}$ percent of your score for this test.

City A currently has a population of 250,000 and City B currently has a population of 350,000. If the population of City A increases at a constant rate of 3% per year and the population of City B increases at a constant rate of 1% per year, then in approximately how many years will the population of the two cities be equal? According to this projection, what will the population of City A be at that time? Explain how you arrived at your answer.

NOTES

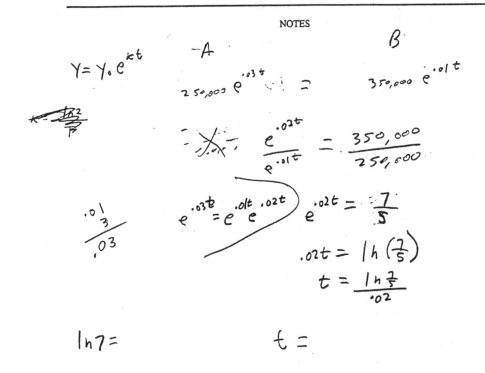

Another analytic solution:

Begin your response to question 3 here.

Population growth can be expressed by the exponential equation $y = y_0 \, e^{kt}$, where y_0 = initial amt., y = final amount, K = rate of growth, and t = time in years.

growth is exponential;

__City A__

$$y = 250,000 \, e^{.03t}$$

__city B__

$$y = 350,000 \, e^{.01t}$$

if we set these two equations equal to each other and solve for t, then we will have the number of years it will take for the population of these two cities to be equal to each other.

populations must be equal;

$$250,000 \, e^{.03t} = 350,000 \, e^{.01t}$$

$$\frac{e^{.03t}}{e^{.01t}} = \frac{350,000}{250,000}$$

$$e^{.02t} = \frac{7}{5} \quad \left(\begin{array}{l} \text{because } e^{.03t} = \\ e^{.02t \, + .01t} \\ = e^{.02t} \, e^{.01t} \end{array} \right)$$

$$.02t = \ln\left(\frac{7}{5}\right)$$

$$t = \boxed{\frac{\ln\left(\frac{7}{5}\right)}{.02}} \text{ years}$$

the time is found (exactly without approximation);

this the answer. I don't have a calculator with me because I thought that the only kind of calculator we could bring was a graphing calculator, and, since I don't have one, I thought I could not bring any. (next page)

(Question 3 continued)

$$t = \frac{\ln\left(\frac{7}{5}\right)}{.02} \text{ years}$$

to find the population of city A at that time,

$$Y = 250,000 \, e^{.03\left(\frac{\ln\frac{7}{5}}{.02}\right)} \longrightarrow t = \text{years.}$$

multiply this out
to get the new population,
$$\underline{Y}.$$

and the answer found and stated
exactly without approximation.

Sample Response 18

Score = 5

Begin your response to question 3 here.

The exhaustive table clearly computes growth as exponential.

CITY A	250,000
YR 1	257,500
2	265,225
3	273,182
4	281,377
5	289,818
6	298,513
7	307,468
8	316,692
9	326,193
10	335,979
11	346,058
12	356,440
13	367,133
14	378,147
15	389,492
16	401,177
* 17	413,212
18	425,608
19	438,376
20	451,528

CITY B	350,000
YR 1	353,500
2	357,035
3	360,605
4	364,211
5	367,853
6	371,532
7	375,247
8	378,999
9	382,789
10	386,617
11	390,483
12	394,388
13	398,332
14	402,315
15	406,338
16	410,401
17	414,505
18	418,650
19	422,836
20	427,065

The time is found when the populations are equal.

(OVER)

(Question 3 continued)

In appoximately 17 years the population of City A and the population of City B will be equal.

The population of City A will be about 413,000.   And finally a population for that time is given as the answer.

By scheduling the annual growth of City A at 3% and City B at 1%, then comparing total population annually, the schedule shows that in year 17 the populations of City A and B are approximately 413,200 and 414,500, respectively, at the end of the year.

The difference of 1,300 (414,500 - 413,200) is immaterial.

Sample Response 19

Score = 3

A graphic approach shows that the growth is exponential, an intersection of the graphs is the solution, but no further progress.

Begin your response to question 3 here.

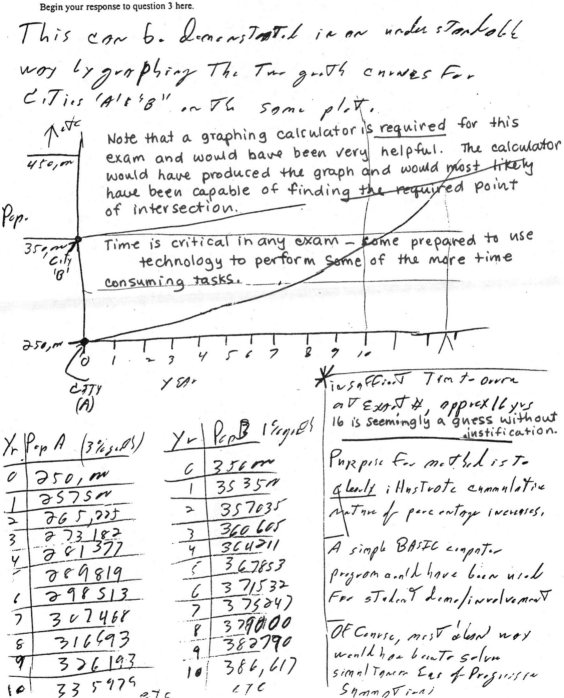

This can be demonstrated in an understable way by graphing the two growth curves for Cities "A" & "B" on the same plot.

Note that a graphing calculator is <u>required</u> for this exam and would have been very helpful. The calculator would have produced the graph and would most likely have been capable of finding the required point of intersection.

Time is critical in any exam — come prepared to use technology to perform some of the more time <u>consuming tasks.</u>

* insufficient Treatment of Exact #, approx 16 yrs
16 is seemingly a guess without <u>justification.</u>

Purpose for method is to clearly illustrate cummulative nature of percentage increases.

A simple BASIC computer program could have been used for student demo/involvement

Of Course, most elegant way would have been to solve simultaneous Eqs of Progressive Summations

Yr	Pop A (3% gr.05)
0	250,000
1	2575000
2	265,225
3	273,182
4	281377
5	289819
6	298513
7	307468
8	316693
9	326193
10	335979 etc

Yr	Pop B 1% gr.05
0	350000
1	353500
2	357035
3	360605
4	364211
5	367853
6	371532
7	375247
8	379000
9	382790
10	386,617 etc

Sample Response 20

Score = 3

Question 3

This question is worth $33\frac{1}{3}$ percent of your score for this test.

City A currently has a population of 250,000 and City B currently has a population of 350,000. If the population of City A increases at a constant rate of 3% per year and the population of City B increases at a constant rate of 1% per year, then in approximately how many years will the population of the two cities be equal? According to this projection, what will the population of City A be at that time? Explain how you arrived at your answer.

NOTES

$$2^x = 8$$

$$8 \log 2 = x$$

$$x$$

$$2^n = 8$$

$$2 \log 4 = n$$

Begin your response to question 3 here.

A) We want to find when both cities population will be equal

City A grows at 3% and City B grows at 1% every year.

n = years   add <u>one</u> to 3% & 1% to get % increase every year.

$250,000 (1.03)^n$ = growth of city A   Population growth is certainly exponential.

$350,000 (1.01)^n$ = growth of city B

now set them equal to each other & solve

A good strategy is shown; the time (n) must be found.

$$250,000 (1.03)^n = 350,000 (1.01)^n$$

This is a 3 without going further.

If the time had been found, 4 points.

and if the population ∧ found, 5 points.
_(had been)

Sample Response 21

Score=2

Begin your response to question 3 here.

A = 250,000 at 3% per yr.

B = 350,000 at 1% per yr.

when will:

$(250,000)(0.03)^t = (350,000)(0.01)^t$ after year 1

$$\frac{7500 =}{250,000 + 7500}$$

$$\begin{array}{r} 257,500 \\ +\ +7725 \\ \hline 265225 \end{array}$$ yr. 2

265,225 + 7957 =
273,182   yr 3

273,182 + 8195 =
281,377   yr. 4

281,377 + 8441 =
289,818   yr. 5

289,818 + 8695 =
298,513

The growth is clearly exponential but nothing else is demonstrated.

298,513 + 8955
307,468

If the data had been organized in a table — and calculations performed on the required calculator, this approach might have been successful.

307,468 + 9224
316692

316,692 + 9501
326192

Sample Response 22

Score = 2

Question 3

This question is worth $33\frac{1}{3}$ percent of your score for this test.

City A currently has a population of 250,000 and City B currently has a population of 350,000. If the population of City A increases at a constant rate of 3% per year and the population of City B increases at a constant rate of 1% per year, then in approximately how many years will the population of the two cities be equal? According to this projection, what will the population of City A be at that time? Explain how you arrived at your answer.

NOTES

$P_A = 250,000 + .03y$

$P_B = 350,000 - .01y$

$250,000 + (250,000 \cdot .03) +$
$250,000 +$

$P_1 \qquad P_1 \cdot .03 +$
$250,000 +$

$P_1 = 250,000$

$P_2 = 250,000 \cdot .03$

$P_3 = P_2 \cdot 0$

$P_1 = 250,000 + (250,000 \cdot .03)$
$P_2 = P_1 + (P_1 \cdot .03)$
$P_3 = P_2 + (P_2 \cdot .03)$

growth is proportional to the population;

1 257500	1 344500
2 265225	2 343035
3 273151	3 337605
4 281377	4 334209
5 289518	5 332847
6 298512	6 329517
7 307467	7 326224
8 316691	8 327962
9 326191	9 319733
10 335976	10 316536

but one population is shown increasing and other decreasing;

so the paper is a 2-point effort.

Begin your response to question 3 here.

THE POPULATION OF CITY A CAN BE DETERMINED FROM THE EQUATION 250,000 + .03y, WHERE y = NUMBER OF YEARS FROM THE CURRENT TIME.

CITY B POPULATION IS 350,000 - .01y

P_A (POPULATION OF CITY A) = 250,000

P_1 (AFTER 1 YEAR) = $P_A \cdot .03 + P_A$

	CITY A	CITY B
1	257,500	346,500
2	265,225	343,035
3	273,181	339,605
4	281,377	336,209
5	289,818	332,847
⋮		
10	335,976	316,536

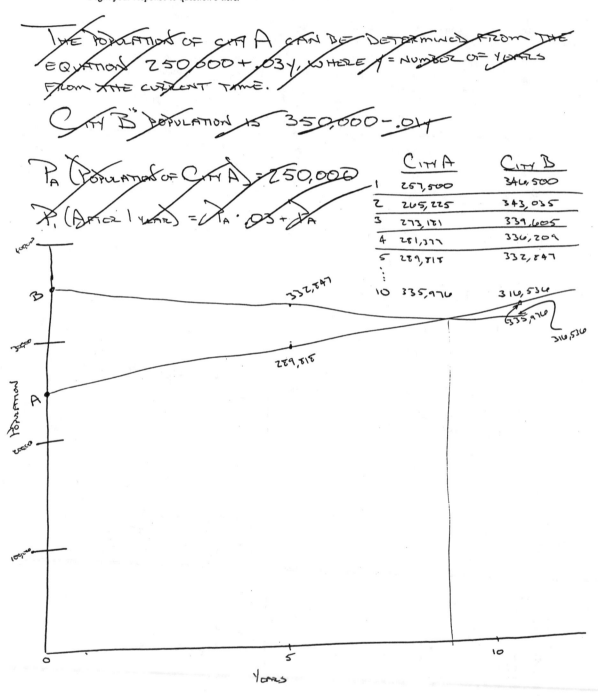

(Question 3 continued)

AFTER GRAPHING THE TWO POPULATIONS THEY ARE EQUAL BETWEEN 8 & 9 YEARS. I READ THIS FROM THE GRAPH AFTER PLOTTING THE POPULATION OF EACH CITY FOR 10 YEARS.

THE POPULATION OF CITY A WILL BE BETWEEN 316,691 & 326,191

Sample Response 23

$$Score = 1$$

Question 3

This question is worth $33\frac{1}{3}$ percent of your score for this test.

City A currently has a population of 250,000 and City B currently has a population of 350,000. If the population of City A increases at a constant rate of 3% per year and the population of City B increases at a constant rate of 1% per year, then in approximately how many years will the population of the two cities be equal? According to this projection, what will the population of City A be at that time? Explain how you arrived at your answer.

NOTES

$t =$ time
$P =$ Population city

$P =$

$$1$$
$$250,000$$
$$.03$$
$$\overline{750000}$$

Begin your response to question 3 here.

$t = time \qquad t = 0 = now$

$P^A =$ Population of City A

$P^B =$ Pop of City B

$$P^A = .03 \overset{7500}{(250,000)} t + 250,000$$

$$P^B = .01 \overset{3500}{(350,000)} t + 350,000$$

The growth is shown to be linear, not exponential — so it is a totally inappropriate approach to the problem.

The populations will be equal when $\Rightarrow$

$$P^A = P^B$$

It is a shame since the populations are set equal, solved for the time, and the population for that time computed,

$$7500t + 250,000 = 3500t + 350,000$$

but all resting on false assumptions about the nature of this problem.

$$\begin{array}{r} 7500t = 350,000 \\ -3500t = -250,000 \\ \hline 4000t = 100,000 \end{array}$$

$$t = \frac{100,000}{4000} \quad 25$$

$$\boxed{t = 25 \, years}$$

An 25 years the populations will be equal.

(Question 3 continued)

In 25 years the population of City A will be

$$P^A = 7500(t) + 250,000$$

$$P^A = 7500(25) + 250,000$$

$$P^A = 187,500 + 250,000$$

$$P^A = 437,500$$

The population of City A will be $\boxed{437,500.}$

$$\begin{array}{r} 7500 \\ \times\ 25 \\ \hline 37500 \\ 150000 \\ \hline 187500 \\ 250000 \\ \hline 437500 \end{array}$$

Question 4

This section begins with a discussion of the solution to Question 4, followed by several actual responses, with comments from the lead scorers to explain how the scoring guide above was used to rate each response.

Question 4–Solution

It is often useful in solving geometry problems to draw line segments that bring out relationships among parts of the geometric figure described or given in the problem statement.

Two important facts about circles are that all radii are of the same length and that tangents meet radii at right angles. This suggests that drawing radii OP and OQ might lead to a way of proving the statement.

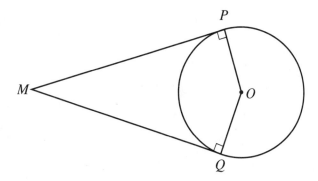

At this point, symmetries in the picture suggest drawing line segment MO. This produces two right triangles with one congruent side ($OP \cong OQ$) and a common side (OM) that makes the two right triangles congruent.

Given this strategy, a proof can be constructed as follows.

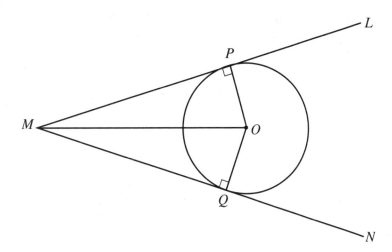

Draw lines PO, OQ, and MO, as shown in the figure above. Note that because P and Q are points of tangency, angles OPM and OQM are right angles. Also, the lengths of OP and OQ are equal since both are radii of the circle. This means that OPM and OQM are congruent right triangles (both have OM as the hypotenuse and $OP \cong OQ$).

Because these two triangles are congruent, the length of MP is equal to that of MQ, since they are corresponding parts of congruent triangles.

Question 4–Sample Responses

Sample Response 24

Score=5

This question is worth $33\frac{1}{3}$ percent of your score for this test.

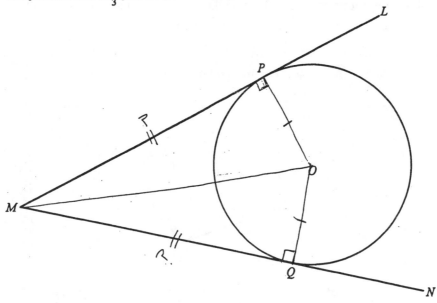

O is the center of the circle.
ML and MN are tangent to the circle at points P and Q, respectively.

Prove that the length of line segment MP is equal to the length of line segment MQ.

___

NOTES

Begin your response to question 4 here.

What is given is clearly set forth.

Given: O is center

ML tangent at P
MN tangent at Q

The proof is a sequence of statements that include justification

PROOF:

Since O is center and P and Q are points on the circle, $\overline{OP} = \overline{OQ}$ by definition of radius.

Since ML is tangent to the circle at P, by definition of tangent $\angle MPO = 90°$.

Right angles are found;

And by the same token $\angle MQO = 90°$. right triangles established;

If we were two draw a line segment from M to O, we form two right triangles, $\triangle MPO$ and $\triangle MQO$

Since $OP = OQ$ and MO is the hypotenuse for each triangle, we know that $\triangle MPO \cong \triangle MQO$ by the hypotenuse leg theorem. the triangles shown congruent;

Since the two triangles are congruent, by definition of congruent $\overline{MP} = \overline{MQ}$ because corresponding parts of congruent triangles are equal.

and corresponding parts of these triangles are the required segments.

Sample Response 25

Score = 5

This question is worth $33\frac{1}{3}$ percent of your score for this test.

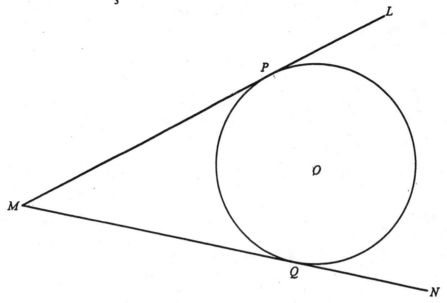

O is the center of the circle.
ML and MN are tangent to the circle at points P and Q, respectively.

Prove that the length of line segment MP is equal to the length of line segment MQ.

NOTES

Begin your response to question 4 here.

First we would draw in the radii from point O to points P and Q. From geometry we know that a tangent drawn to a circle has a point of tangency perpendicular to the radius at their point a intersection. That is the point of tangency is perpendicular to the radius at their point of intersection. So we have the following

A more analytic approach using the Pythagorean Theorem rather than congruence.

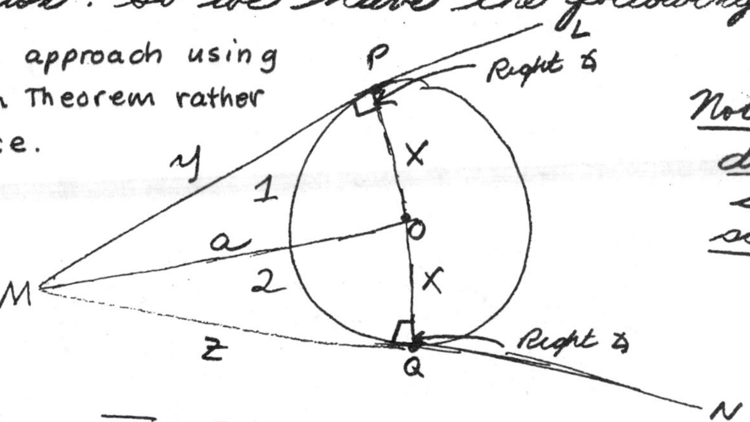

Not drawn to scale

More info ↓

We know $\overline{PO} = \overline{PQ}$ since they are both radii of the same circle. Call them x. We want to prove that $\overline{MP} = \overline{MQ}$. Call $\overline{MP} = y$ and $\overline{MQ} = z$. We want to then prove $y = z$. Call $\overline{MO} = a$. Using the (Right Δ's) pythagorean theorem Δ1 $y^2 + x^2 = a^2$. So $y^2 = a^2 - x$ So $y = \sqrt{a^2 - x^2}$. Again using the theorem Δ2 we have $z^2 + x^2 = a^2$. So $z^2 = a^2 - x^2$. So $z = \sqrt{a^2 - x^2}$ Hence $y = z$. **Thus $\overline{MP} = \overline{MQ}$**

Sample Response 26

$Score = 4$

Begin your response to question 4 here.

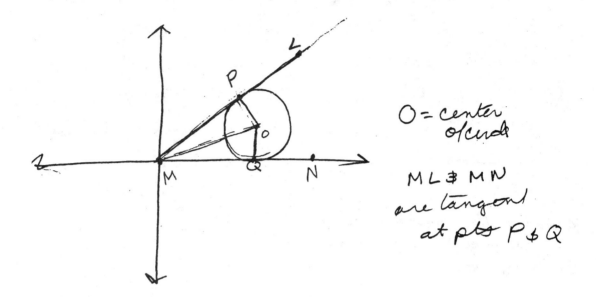

O = center of circle

ML & MN are tangent at pts P & Q

Segment $OP \equiv OQ$ because the are both radii of the circle, thus by def of a radius and a circle $OP = OQ$

A discussion of tangents is extraneous.

If we draw in a segment MO we could create two triangles with the same tangent value

$MO = MO$
$OP = OQ$

Discussion turns to the Pythagorean Theorem
and is fine, except at no point is it established
that there is a right angle.

(Question 4 continued)

by the pythagorean theorem
the sides of these right
triangles with Hypotenus MO
would be for

MOP

$MP^2 + OP^2 = MO^2$ and MOQ

$$MQ^2 + OQ^2 = MO^2$$

so $MP^2 + OP^2 = MQ^2 + OQ^2$

$$OP = OQ$$

$$MP^2 = MQ^2$$

$$MP = MQ$$

The lack of justification for a right angle
is a definite flaw — but argument is
otherwise successful. So the solution
is given 4 points.

Sample Response 27

Score=4

Question 4

This question is worth $33\frac{1}{3}$ percent of your score for this test.

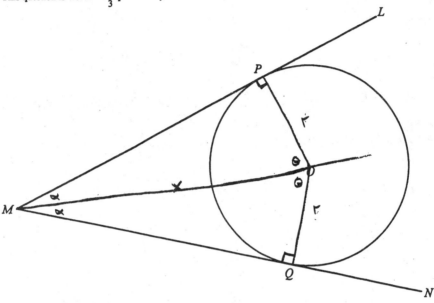

O is the center of the circle.

ML and MN are tangent to the circle at points P and Q, respectively.

Prove that the length of line segment MP is equal to the length of line segment MQ.

NOTES

Begin your response to question 4 here.

Since ML & MN are tangent to the circle O, the bisector of ∠LMN must pass through the center of the circle (Pt. O). Since $\vec{MO}$ is the bisector, m∠PMO = m∠QMO (lets call it α). Also, m∠MPO = m∠MQO = 90° by definition of tangent. Also OP = OQ since they are both radii. (lets call it r).

∠POM = 180 - 90 - α

∠QOM = 180 - 90 - α

} 3 ∠'s add to 180°

so m∠POM = m∠QOM   (let call it θ)

We can then use the Angle-Side-Angle Theorem to show that △OPM ≅ △OQM.

It follows that PM ≅ QM   [or mP ≅ mQ] which is what we were to prove.

There is no justification for the assertion that the angle bisector passes through the center of the circle.   So the solution is scored 4 points.

Sample Response 28

Score = 3

Question 4

This question is worth $33\frac{1}{3}$ percent of your score for this test.

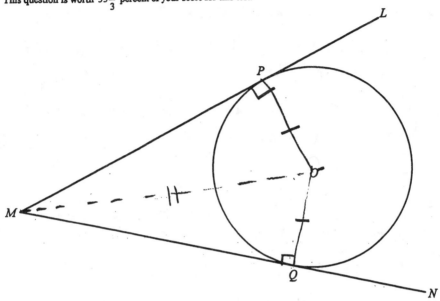

O is the center of the circle.
ML and MN are tangent to the circle at points P and Q, respectively.

Prove that the length of line segment MP is equal to the length of line segment MQ.

NOTES

Begin your response to question 4 here.

$\overline{OQ} \cong \overline{OP}$ because the radius of a circle are congruent. $\angle MQO \cong \angle MPO$ is given because $\overline{MP}$ and $\overline{MQ}$ are tangent. $\overline{MO}$ bisects $\angle PMQ$. And $\overline{MO} \cong \overline{MO}$ reflexively. $\triangle MQO \cong \triangle MPO$ by side angle side. Therefore $\overline{MP} \cong \overline{MQ}$ by corresponding parts of corresponding triangle are congruent and $MP = MQ$

There is substantial progress, but $\angle PMQ$ is asserted to be bisected without justification, and "side-angle-side" asserted with the wrong angle.

With a little more care about organization, this might have been better than 3 points.

Sample Response 29

Score = 2

This question is worth $33\frac{1}{3}$ percent of your score for this test.

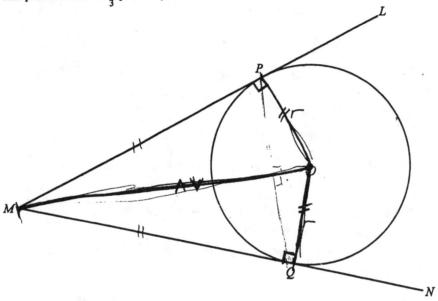

O is the center of the circle.
ML and MN are tangent to the circle at points P and Q, respectively.

Prove that the length of line segment MP is equal to the length of line segment MQ.

NOTES

$$a^2 + b^2 = c^2$$

Begin your response to question 4 here.

Proof	Reason
PO = QO	Radii of the same circle are equal
MO = OM	Same line
MP = MQ	If two sides of two different triangle are conguent, then the third sides must be equal.

Has completed the figure with useful auxilary lines, noted equal radii and a common line, and seems to know where one needs to go, but lacks the understanding of congruence necessary to make further progress. A score of 2 points.

mp is equal to MQ because all radius of one circle are equal, and MO in both triangles is equal because it is the same line, therefor the third side of each △ must be equal.

Sample Response 30

Score = 1

Begin your response to question 4 here.

Since we have a circle with center O, we know that every point on the circle is the same distance from the center. We know that point P & point Q are the same distance from point O. We also know that given a point M some place out side the circle that if drawing tangents to the circle from that point m that they will touch the circle in only place on the circle respectfully P & Q. Since we started from the same point and created tangents we know that the points of tangence is equal distance the place of origin therefor segment MP is equal to MQ.

The first sentence is promising — but the next sentence is a restatement of the problem given as an answer. Work is never really begun our the actual problem.

Chapter 11

Solutions and Sample Responses for the
Mathematics: Pedagogy Test and How They Were Scored

► ► ► ► ► ► ► ► ► ► ► ►

This chapter presents actual sample responses to the questions in the practice test (chapter 8) and explanations for the scores they received.

As discussed in chapter 5, each question on the *Pedagogy* test is scored on a scale from 0 to 5. The general scoring guide used to score these questions is reprinted here for your convenience.

Scoring Guide

A response that does not demonstrate an understanding of the mathematics to be presented CANNOT receive a score greater than 2, regardless of any other criteria for higher scores that the response might meet.

<u>Score</u>	<u>Comment</u>
5	■ Clearly demonstrates an understanding of the mathematics to be presented ■ Clearly explains how to present the mathematics to students in a way that is likely to achieve the desired goal(s) ■ Gives a clear and complete response, develops the mathematics in a way that is well motivated (i.e., students can clearly see why the mathematics being presented is worth studying and/or can see the mathematics as the logical consequence of previously studied mathematics)
4	■ Clearly demonstrates an understanding of the mathematics to be presented but may have a notational error or minor mathematical misstatement ■ Explains how to present the mathematics to students in a way that can reasonably be expected to achieve the desired goal(s) ■ EITHER gives an almost complete response and a well-motivated development of the material OR gives a complete response and a fairly well motivated development of the material
3	■ Demonstrates an understanding of the mathematics to be presented ■ Indicates how to present the mathematics to students in a way that can reasonably be expected to achieve the desired goal(s) ■ EITHER gives an almost complete response and a well-motivated development of the material OR gives a complete response and a fairly well motivated development of the material
2	■ EITHER demonstrates a limited understanding of the mathematics to be presented (and may or may not indicate how to teach the mathematics to students in a way that is likely to achieve the desired goal(s)) OR demonstrates an understanding of the mathematics but gives little or no indication of how to present the mathematics to students in a way that is likely to achieve the desired goal(s) ■ Gives an unclear and incomplete response
1	■ EITHER demonstrates a very limited understanding of the mathematics to be presented OR fails to discuss the mathematics at all
0	■ Blank, almost blank, or off topic

Question 1

This section begins with a discussion of the solution to Question 1, followed by several actual responses, with comments from the lead scorers explaining how the scoring guide above was used to rate each response.

Question 1–Solution

To devise an appropriate homework assignment, you must take into account the skills and concepts that the students are likely to have learned (and be in the process of learning) when the homework assignment is given. In this case, the homework should include skills and concepts that would have been covered in a unit on solving quadratic equations that included (i) taking square roots, (ii) factoring, and (iii) solving quadratic equations by completing the square.

A response requires five problems that address concepts involving the solution of quadratic equations. These problems should represent different skills, and all three methods of solving a quadratic equation should be covered in the set of five problems. In addition, in order for the problems presented to be well motivated pedagogically, the problems should show an appropriate progression of concepts and difficulty.

A reasonable approach to developing a response to this question is to think of five different skills and concepts that are covered in the three methods and develop problems that will require the students to use each of the skills and concepts. Typically, this material will be taught by using quadratics that factor fairly easily and will not involve complex numbers.

The following is a list of skills and concepts that could be taught as part of a unit on solving quadratic equations, along with a sample problem for each. (This list does not include all possible skills and concepts that could be taught in the unit.)

Skill: Taking Square Roots

Sample Problem: Solve the equation $x^2 = 25$.

Rationale: This problem reviews material from earlier lessons and will help identify those students who do not consider both the positive and negative square roots when taking the square root to solve a simple quadratic.

Skill: Recognizing the square of a binomial

Sample Problem: Solve the equation $x^2 + 6x + 9 = 0$ by factoring.

Rationale: This problem is a review of factoring and leads to completing the square by developing recognition of squares of binomials.

Skill: Recognizing that quadratics of the form $x^2 + (a+b)x + ab$ can be factored into $(x+a)(x+b)$

Sample Problem: Factor $x^2 + 5x + 4 = 0$.

Rationale: This problem reviews factoring and can be approached again later from the point of view of completing the square to show that when two different methods can be used to solve the equation, both will give the same answer.

Skill: Factoring expressions involving the difference of two squares (i.e., expressions of the form $x^2 - a^2$)

Sample Problem: Solve the equation $x^2 - 36 = 0$ by factoring and using the method of taking square roots.

Rationale: This problem reviews another familiar factoring technique and also shows that the quadratic can often be solved in more than one way.

Skill: Factoring a constant that occurs in every term of an expression

Sample Problem: Solve the equation $3x^2 + 6x + 3 = 0$.

Rationale: This problem involves a slightly more complex form of factoring and prepares the student to factor out the first term of the quadratic, a technique that is often necessary when completing the square.

Skill: Given some details about the coefficients of a quadratic equation, being able to determine the relationship between the coefficients

Sample Problem: If $x^2 + bx + 10$ is of the form $(x + a)^2$, what are all of the values of b between -12 and 12 that will satisfy the equation $x^2 + bx + 10 = 0$?

Rationale: This problem will be more of a challenge and will help the student recognize the relationship that must exist between the coefficient of x and the constant term in order for the quadratic to be a perfect square.

Skill: Given a quadratic equation and the "completed square," determine the constant needed to get from one to the other

Sample Problem: If $x^2 + 12x + 12 = (x + 6)^2 + k$, then find k.

Rationale: This problem is good practice for completing the square.

Skill: Solving an equation by completing the square

Sample Problem: Solve the equation $x^2 + 8x + 7 = 0$ by completing the square.

Rationale: With this problem, the student practices completing the square and sees a problem that is not easily solved by factoring.

Skill: Using an example to conjecture relationships between quadratic equations in "completed square" form and "trinomial" form

Sample Problem: Using the equation $49x^2 - 70x + 25 = (7x - 5)^2 = 0$ as a guide, analyze the equation $ax^2 - bx + c^2 = (kx - c)^2$ and express both k and b in terms of a and c.

Rationale: This problem requires the student to think carefully about what relationships should exist between an equation in "completed square" form and "trinomial" form.

Skill: Rearranging terms and factoring coefficients before completing the square

Sample Problem: Solve the equation $6t - 1 = 2t^2$.

Rationale: This problem will challenge the best students, since the equation must be rearranged and divided through by 2 before proceeding.

Question 1–Sample Responses

Sample 1: Score of 5

The response that begins on the next page clearly demonstrates an understanding of all the mathematics to be presented. The problems show an appropriate progression from easier questions on familiar material to harder questions on new material. All of the problems have only real solutions, which is appropriate for a class just learning how to solve quadratic equations. It is also appropriate that Problem #3 is used to show that some quadratic equations can be solved by more than one of these methods.

The response is clear and complete. A thorough rationale is provided for each of the questions given, and each of the questions is appropriate to the rationale given for that question.

The homework set is one that is likely to achieve the desired goals of reviewing older material along with gaining experience with the new technique of completing the square.

MATHEMATICS: PEDAGOGY
Time—60 minutes
3 Questions

Question 1

1. You are teaching a unit on solving quadratic equations. You have already taught the students how to solve quadratics by taking square roots and by factoring. In your next lesson you plan to teach the students how to solve quadratic equations by completing the square.

   Design a homework assignment for your students to complete after that next lesson. The homework assignment should consist of 5 problems that review previously taught skills and concepts while also providing practice in the newly introduced material.

   Briefly explain your rationale for including the skills and concepts that the problems illustrate.

NOTES

$$x^2 - 4 = 0$$
$$x = \pm\sqrt{4}$$
$$x = \pm 2$$

$$2x^2 - 3 = 0$$
$$x = \pm\sqrt{3}$$
$$x = \pm\sqrt{\tfrac{3}{2}}$$

Factoring.
$$x^2 - 4x + 4 = 0$$
$$(x - 2)^2 = 0$$
$$x = 2$$

$$x^2 - 6x + 9 = 0$$
$$(x - 3)^2 = 0$$
$$x^2 - 5x + 6 = 0$$
$$x - 5x + \left(\tfrac{5}{2}\right)^2 - \left(\tfrac{5}{2}\right)^2 + 6 = 0$$
$$\left(x - \tfrac{5}{2}\right)^2 - \tfrac{25}{4} + 6 = 0$$
$$-\tfrac{25}{4} + \tfrac{24}{4} = 0$$
$$y = \tfrac{5}{2} \pm \sqrt{\tfrac{1}{4}}$$
$$x = \tfrac{5 + 1}{2} \qquad \tfrac{5 - 1}{2}$$
$$x = 3, 2$$

$$3x^2 - 12x + 10 = 0$$
$$x^2 - 4x + \tfrac{10}{3} = 0$$
$$x^2 - 4x + 4 - 4 + \tfrac{10}{3} = 0$$
$$x^2 - 4x + 4$$
$$(x - 2)^2 - \tfrac{12 + 10}{3} = 0$$
$$(x - 2)^2 = \tfrac{2}{3}$$
$$\sqrt{(x - 2)^2} = \pm\sqrt{\tfrac{2}{3}}$$
$$x = 2 \pm \sqrt{\tfrac{2}{3}}$$

Begin your response to question 1 here.

I will assign them these 5 problems:

1) $x^2 - 4 = 0$
2) $2x^2 - 3 = 0$
3) $x^2 - 4x + 4 = 0$
4) $x^2 - 5x + 6 = 0$
5) $3x^2 - 12x + 10 = 0$.

The first problem $x^2 - 4 = 0$ is the review on how to solve quadrat og taking square roots: $x^2 - 4 = 0$

move (-4) to the other side $\quad x^2 = 4$.
square root both sides $\quad \sqrt{x^2} = \sqrt{4}$
put $\pm$ sign on $\sqrt{4}$ $\quad x = \pm\sqrt{4}$
Solve for $\sqrt{4}$ $\quad x = \pm 2$.

I need this problem so I can remind them the material I have taught several days ago.

The second problem is like the first, the difference is that there is a constant in front of x^2 : $2x^2 - 3 = 0$.
This problem is included to see if students still remember that they need a constant 1 as leading coefficient. They will need to divide everything by 2. : $x^2 - \frac{3}{2} = 0$.
then the steps on number 1 followed $\quad x = \pm\sqrt{\frac{3}{2}}$. or $x = \pm\frac{\sqrt{6}}{2}$

Since I have also taught them how to factor to obtain solution of quadratic equations, I include problem #3 : $x^2 - 4x + 4 = 0$ this is a perfect square. And they can easily factor it $(x-2)^2 = 0$ they will obtain $x = 2$ as their solutions (multiplicity 2).
At this time I will ask the students to use completing the square method to solve this problem (3). Since they know already that the answer is 2 then when they do the problem, they will have something to check their work.

Problem number four is a little harder than the previous one. Student may factor it into $(x-3)(x-2)=0$, $x=2,3$. Again I will ask the students to use completing the square after they have factored and obtained solutions for ~~the~~ problem #4.

Since they know that the answer is either 2 or 3, they will be able to check their answers after they have done completing the square As you see $x^2 - 5x + 6 = 0$ when it is completed will have the form $$\left(x - \frac{5}{2}\right)^2 = \frac{1}{4}.$$

Students will then remember to take the square roots of both sides because they have just done this in problem # 1 and 2.

they will then have $x = \frac{5}{2} \pm \frac{1}{2}$ which is the same as $x=3$.

The fifth problem is a take off from problem # 2. I want to remind them that to complete the square, the leading coefficient in this case 3 has to be turned into 1.

By dividing the entire equation by 3 we will have $x^2 - 4x + \frac{10}{3} = 0$ They will then proceed to complete the square because it would be hard to solve the problem using factoring. I included this problem so that students know that they can't depend on one way of solving a particular problem. These three methods (square root factoring and completing the squares) are tools to solve a quadratic equations; One tool may not ~~so be~~ be used to solve all quadratic equations effectively.

Now that they have learned how to incorporate the three tools, they are ready to solve most of quadratic equations.

Sample 2: Score of 4

The response that begins on the next page clearly demonstrates an understanding of the mathematics to be presented. Although the response does not directly show a solution by taking square roots, factoring, or completing the square, it is clear from the problems selected and the rationale given for each problem that the examinee knows which method is most appropriate to each problem and thus demonstrates a clear understanding of the mathematics.

The response is clear and complete, and an appropriate rationale is given for each response.

The development of the material is only "fairly well motivated" in that two of the homework problems (#3 and #4) have complex solutions. This is not entirely appropriate for the students learning the method of completing the square for the first time. The connection made in #5 between two of the methods does serve as good motivation of the material. Except for the complex solutions, the homework problems are likely to produce the desired goals of review of familiar techniques along with practice on the new method.

1. $(1 + 2)^2 = 9$

I would include this problem because the students should realize that the problem is easily completed by the square root method. This problem would be used as a review example.

2. $x^2 + 8x + 16 = 0$

I would include this problem because I want the students to continue practicing factoring. They should see that this equation is easily factored because the first and last terms are squared terms.

3. $x^2 + 4x + 7 = 0$
4. $2x^2 + 8x + 20 = 0$

These two problems would allow the students to practice the completing the square method that they have just learned. #3 is a fairly easy and strait foward problem. #4 is a little more complicated because the students have to remember that the leading coefficient must be 1 so the students might divide through by

GO ON TO THE NEXT PAGE.

−2−

2 before they begin completing the square.

5. $x^2 + 7x + 12 = 0$

I would ask the students to complete the problem by factoring and by completing the square. I would also have them tell which way they like best $\underset{\wedge}{\text{& why}}$. This helps the students see that a problem can be solved in more than one way.

GO ON TO THE NEXT PAGE.

Sample 3: Score of 3

The response that begins on the next page demonstrates an understanding of the mathematics to be presented. The response demonstrates an awareness of which problems require completing the square and which ones can be done by factoring or taking square roots. The first steps in completing the square are shown on the notes page, further demonstrating an understanding of the mathematics. This understanding is less clearly demonstrated than on the responses that received higher scores.

The problems do not show an appropriate progression in difficulty or in the techniques required to solve them. Also, some of the rationales do not fit the problem given. For example, the rationale for Problem #2 says that completing the square is necessary for this problem, but factoring would be a much easier way to solve the equation. Also the rationale for Problem #5 indicates that this will be a "different looking problem" but this problem looks much like Problem #2. These inappropriate rationales make the response fairly well motivated at best and make the examinee's understanding of mathematics less clear. The motivation of the response is improved by connections that are suggested between solving by factoring and solving by completing the square in Problem #1. Thus this response is complete and fairly well motivated.

Question 1

You are teaching a unit on solving quadratic equations. You have already taught the students how to solve quadratics by taking square roots and by factoring. In your next lesson you plan to teach the students how to solve quadratic equations by completing the square.

Design a homework assignment for your students to complete after that next lesson. The homework assignment should consist of 5 problems that review previously taught skills and concepts while also providing practice in the newly introduced material.

Briefly explain your rationale for including the skills and concepts that the problems illustrate.

NOTES

1. $3x^2 + 4x + 4 = 0$

2. $x^2 + 5x = 0$

3. $x^2 = 16$

4. $4x^2 + 3x + 1 = 0$

5. $-x^2 + 2x = 0$

$$ax^2 + bx + c = 0$$

$$x^2 + \frac{b}{a}x = -\frac{c}{a}$$

1q-00d-lmi

GO ON TO THE NEXT PAGE.

-2-

Begin your response here.

The first problem deals with a review of solving quadratic equations by factoring, $x^2 + 4x + 4 = 0$. This is included to gain knowledge of how a student would see this problem after a new concept was learned. This problem would be done more easiley by factoring than completing the square. It would be important for the student to know the best way to solve these equation.

The second problem would give an easy completing the squares problem. This would be done to build confidence and hopefully provide easy recognition that $x^2 + 5x = 0$ needs the square to be completed.

$x^2 = 16$ would be a review for the student.

$4x^2 + 3x + 1 = 0$ would require the student to work hard to solve this problem by completing the squares and for a different looking problem $-x^2 + 2x = 0$ will be given.

GO ON TO THE NEXT PAGE.

Sample 4: Score of 2

The response that begins on the next page demonstrates only a limited understanding of the mathematics to be presented. The problems chosen and the questions asked fail to address appropriately the method of taking square roots, a method closely related to completing the square. Since all of the questions but the last one are solvable by factoring, it is unclear whether this examinee understands the method of completing the square when that is the appropriate approach. The fact that the last problem is the only one that requires completing the square does suggest that the examinee may know when completing the square is required, but the examinee does not mention that this is the case or that the first four problems can be done by either method. The examinee's comments at the end of the response make it less clear that the examinee understands that there are problems for which completing the square is required.

The motivation of the problems falls short in three ways. There is no progression from the method of square roots to other methods; the fourth question asked effectively answers the third question from the students' point of view; and there is no specific explanation of why the problems given were selected for homework.

This response is unclear and incomplete compared to those that received a score greater than 2.

MATHEMATICS: PEDAGOGY

Time—60 minutes
3 Questions

Question 1

1. You are teaching a unit on solving quadratic equations. You have already taught the students how to solve quadratics by taking square roots and by factoring. In your next lesson you plan to teach the students how to solve quadratic equations by completing the square.

Design a homework assignment for your students to complete after that next lesson. The homework assignment should consist of 5 problems that review previously taught skills and concepts while also providing practice in the newly introduced material.

Briefly explain your rationale for including the skills and concepts that the problems illustrate.

NOTES

1) Factoring
2) Completing the Square

   HW assignment.

    a) $x^2 + 5x + 6 = 0$
    b) $x^2 + 3x - 4 = 0$
    c) $x^2 + 4x - 4 = 0$
    d) $2x^2 + 4x - 6 = 0$
    e) $3x^2 + 5x + 7 = 0$

3)

$(2x + 2)(x - 3)$

$(3x +)(x +)$

Begin your response to question 1 here.

I would begin my lesson plan by discussing the previous lessons for solving quadratic equations and discuss any problems or questions the students have on solving quadratics by taking square roots and factoring.

We would then begin in on our next lesson on completing the square. I would be sure to tell the students that it is very possible that you may be able to factor the equation, but we are now going to solve our quadratic equations by completing the square. After I present the material on this method of solving quadratics, I will give the students five problems for homework.

I would say tonight's homework consist of five problems. They are

1) $x^2 + 5x + 6 = 0$
2) $x^2 + 3x - 4 = 0$
3) $x^2 + 4x - 4 = 0$
4) $2x^2 + 4x - 6 = 0$
5) $3x^2 + 5x + 7 = 0$.

There are several different items I would like for you to tell me about each problem.

① Can this quadratic be solved by taking square roots?

② Can this quadratic be solved by factoring?

③ Can the quadratic be solved by completing the square?

(Question 1 continued)

④ Solve each of the quadratic equations by completing the square.

The purpose of getting the students to figure out how to solve the quadratic equations is for them to realize there is not just one way of solving quadratics. They will understand that they will have a choice as to which way they prefer to solve these equations on a quiz and on a test.

Sample 5: Score of 1

The response that begins on the next page demonstrates a very limited understanding of the mathematics to be presented. The expressions presented are quadratic, but none are equations. The examples chosen to review the methods of taking square roots and factoring are not solvable by those methods without completing the square or using the quadratic formula. The work shown for solving one of the problems is incorrect and clearly demonstrates a lack of understanding of the mathematics.

Begin your response to question 1 here.

Before I start teaching the concept, review how to solve quadratic by taking a square roots and by factoring. This review should relate and recall their background knowledge to the new concept, "Completing the square".

Write on the board an example problem on the board such as $2x^2 + 2x + 4$ and call on the student to solve for x. The students' response should explain how they would do it (Factor or find the square root). Use another example such as $3x^2 + 3x + 5$.

Introduce Completing the Square by telling them that "this" is another method that they can use to solve for x.

By using Direct Instruction, the students should be motivated by relating the. Completing of the square to the completing of a parcel in finding a clue to get the missing treasure.

Write a sample problem

Steps ① Work with first 2 numbers
② The missing number is the dividing of the second number by 2 ($\frac{2x}{2} = 1$) But add that same 1 to the 4

$$x^2 + 2x + 4.$$
$$(x^2 + 2x + \underline{1}) + 4$$
$$(x^2 + 2x + 1) + 4 + 1$$
$$(x^2 + 2x + 1) + 5$$

(Question 1 continued)

$$(x^2 + 2x + 1) + 5$$

Factor: $(x+1)(x+1) + 5$

Set $(x+1) = 0$ and
    Solve for x.

$x + 1 = 0$
$\quad -1 \quad -1$
_____
$x + 0 = -1$
$\boxed{x = -1}$

$x + 1 = 0, \quad x + 1 = 0$
$\underline{\quad -1 \quad -1} \quad \underline{\quad -1 \quad -1}$
$x = -1 \qquad x = -1$

Home work: ① $x^2 + 4x + 3$

② $x^2 + 8x + 4$

③ $2x^2 + 4x + 4$

④ $3x^2 + 9x + 3$

⑤ $3x^2 + 6x + 2$

Question 2

This section begins with a discussion of the solution to Question 2, followed by several actual responses, with comments from the lead scorers explaining how the scoring guide above was used to rate each response.

Question 2–Solution

The students described in the question are having trouble with equivalent fractions, although they have already been exposed to the topic. A strategy using pictures or manipulatives uses concrete objects to help students who have trouble with the abstract concept of equivalent fractions to understand the concept in a concrete way. In the long run, however, using pictures or manipulatives to determine when *any* two fractions are equivalent would be cumbersome and time-consuming. Therefore, it is appropriate that the strategy begin with extensive use of manipulatives or pictures so that students develop a firm concrete understanding of what it really means for two fractions to be equivalent. The lesson should then progress from the manipulative examples to develop a general method for determining without manipulatives whether any two fractions are equivalent.

Question 2–Sample Responses

Sample 6: Score of 5

The response that begins on the next page clearly demonstrates an understanding of the mathematics to be presented. The math is presented to the students in a way that is likely to achieve the two desired goals: understanding the meaning of equivalent fractions and developing a method to determine whether any two fractions are equivalent. Multiple examples are given, using a pie chart to show that the idea of equivalence is valid for different types of fractions, and the general method shown for determining equivalence of fractions is one method that will work for any pair of fractions.

The response is clear and complete. The response is well motivated in that it moves from specific visual examples to a general abstract method and all of the examples given are easy to understand.

Question 2

Begin your response here.

① I would take a pie-chart and begin by showing the pictorial representation of $\frac{1}{2}$.

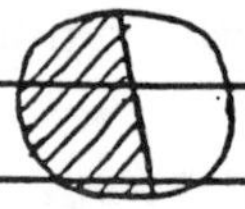

② I would then show the following graph of $\frac{2}{4}$ and preserve the picture above by super-imposing the graph of $\frac{2}{4}$ on to it

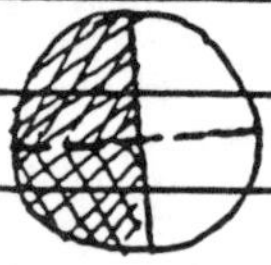

Then $\frac{3}{6}$:

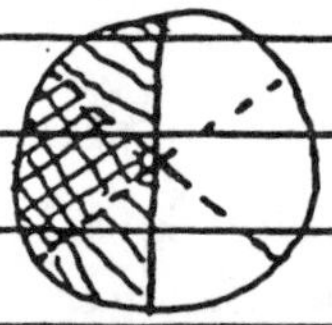

I would then note that rather we take $\frac{1}{2}$, $\frac{2}{4}$, or $\frac{3}{6}$ of the circle we still are considering the same area expressed by the three ratios.

② The same could then be performed use $\frac{1}{3}$, $\frac{2}{6}$, $\frac{3}{9}$

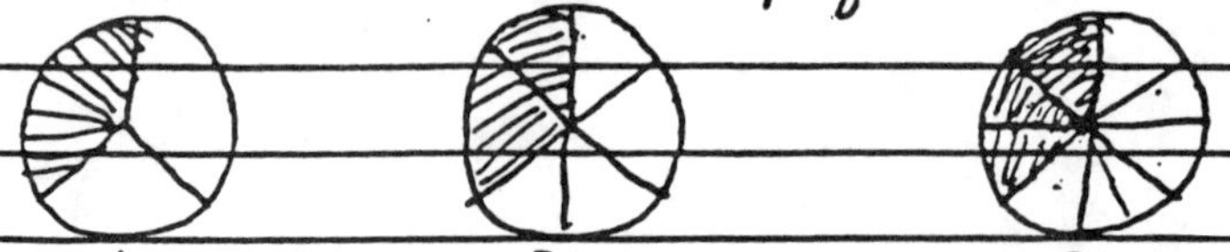

$\frac{1}{3}$ $\frac{2}{6}$ $\frac{3}{9}$

GO ON TO THE NEXT PAGE.

The objective by these pictures would be to see that equivalent areas are covered regardless of the expression of the ratio thus making the ratios equivalent.

③ Fraction can be shown to equivalent by cross-multiplying the numerator and the denominator. Example

$$\frac{1}{5} \overset{?}{=} \frac{2}{10}$$

If we cross multiply the denominators and numerators and their resulting products are equal then the fractions are equal.

$$1 \cdot 10 = 10 \qquad 2 \cdot 5 = 10$$

$$\text{So} \quad \frac{1}{5} = \frac{2}{10}$$

Example:

$$\frac{1}{4} \overset{?}{=} \frac{3}{6} \qquad 1 \cdot 6 = 6 \qquad 3 \cdot 4 = 12$$

$$\text{So} \quad \frac{1}{4} \neq \frac{3}{6}$$

GO ON TO THE NEXT PAGE.

This works because if the resulting products are the same then one fraction is a multiple of the other and value of the ratio is preserved.

Sample 7: Score of 4

The response that begins on the next page clearly demonstrates an understanding of the mathematics to be presented. The portion of the response dealing with pictures or manipulatives demonstrates how to present that material to the students in a way that is likely to achieve the goal of understanding the meaning of equivalent fractions.

The response is complete, but the portion of the response dealing with the development of a general method to determine whether two fractions are equivalent is only fairly well motivated. Though the method described can technically be implemented for any pair of fractions, it will be very difficult for students to use, especially when neither fraction is in lowest terms (e.g., to determine the equivalence of $\frac{3}{6}$ and $\frac{4}{8}$). The method is carefully explained and is likely to begin developing the ability to determine whether two fractions are equivalent.

-2-

Begin your response here.

We could begin by at with the following picture and explain that the circle is divided into two equal parts.

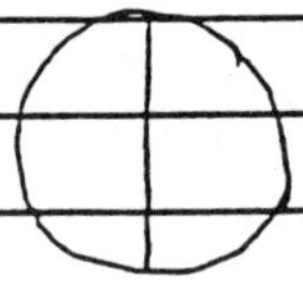

If we take one of those parts and shade it in, the following picture results

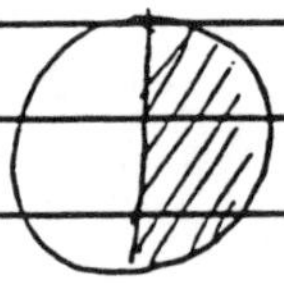

We have shaded 1/2 of the circle

Now let's take a circle and divide it into four equal parts:

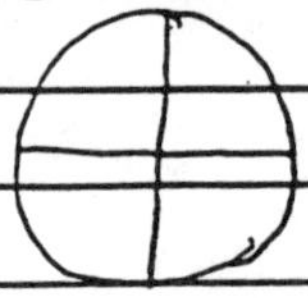

If we shade 2 parts of the above circle our picture will look like the following:

We have shaded 2/4 of the circle.

GO ON TO THE NEXT PAGE.

–3–

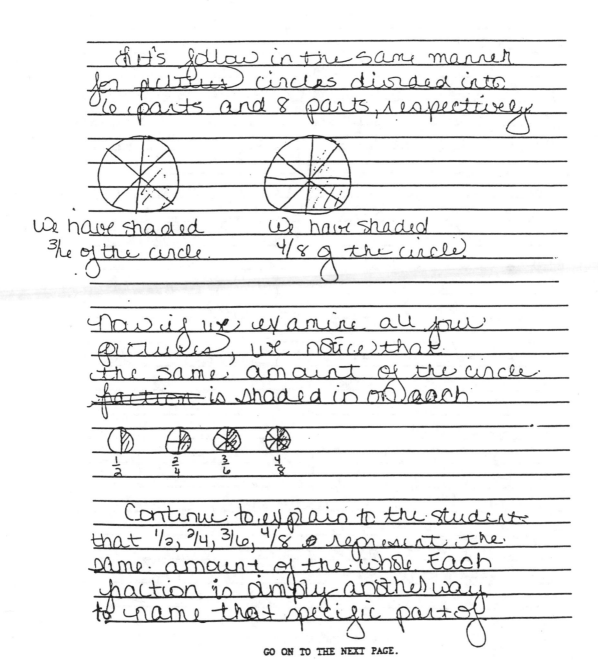

it's follow in the same manner for [crossed out] circles divided into 6 parts and 8 parts, respectively.

We have shaded 3/6 of the circle.

We have shaded 4/8 of the circle.

Now if we examine all four examples, we notice that the same amount of the circle [fraction crossed out] is shaded in [or] each

$\frac{1}{2}$ $\frac{2}{4}$ $\frac{3}{6}$ $\frac{4}{8}$

Continue to explain to the students that 1/2, 2/4, 3/6, 4/8 represent the same amount of the whole. Each fraction is simply another way to name that specific part of

GO ON TO THE NEXT PAGE.

-4-

the whole.

After the students understand the conceptual part of equivalent fractions, explain the numerical manipulation that is involved.

Explain that we have $1/2$ and if we multiply it by any number that equals 1, we will get the same number. (This idea as with the others can be ~~used~~ implemented through questioning techniques rather than mere lecture.) For example,

$1/2 \times 1 = 1/2$

$1/2 \times 2/2 = 2/4$ (~~Since we~~ Remember that we know $2/2 = 1$)

$1/2 \times 3/3 = 3/6$

$1/2 \times 4/4 = 4/8$.

Thus, we have seen that if we multiply $1/2$ by 1 ~~for any~~, ~~then~~ we get the same number. That is,

$1/2 = 2/4 = 3/6 = 4/8$.

Continue to explain that if we started with a fraction, say $5/10$, we see that there is 1

GO ON TO THE NEXT PAGE.

–5–

5 in the numerator and
2 5's in the denominator—
which reduces the fraction)
to ½. We know, then, that
5/10 is equivalent to ½ because
we can multiply ½ by ($(5/5)$
to get 5/10 and because
we can draw a diagram:

½   5/10

We see that ½ and 5/10 are
the same part of the whole
circle.
        Now have the students
give some examples of their own.
Have them draw pictures and
name at least 3 fractions that
are equivalent to the one they posed.

GO ON TO THE NEXT PAGE.

Sample 8: Score of 3

The response that begins on the next page demonstrates an understanding of the mathematics to be presented. By showing just one manipulative example and only briefly discussing a method for determining whether two fractions are equivalent, the examinee is only indicating (rather than demonstrating with multiple examples and more thorough discussion) how to present the mathematics in a way that can reasonably be expected to achieve the desired goals.

The examples presented are well motivated. The small number of examples, along with the lack of explanation for the portion of the response dealing with determining whether two fractions are equivalent, renders the response almost incomplete rather than complete.

Question 2

2. A small group of students in your seventh-grade math class is unable to determine whether two fractions are equivalent. Describe a strategy, using pictures or manipulatives, that you could use to help foster the students' conceptual understanding of equivalent fractions. Your strategy should stress understanding of what it means for fractions to be equivalent and the development of the ability to determine whether fractions are equivalent.

Response begins here:

One strategy in helping the students understand would be to include diagrams with the fractions, particularly for the smaller fractions. For example, if they had difficulty seeing that $\frac{1}{3} = \frac{2}{6}$, I would draw two equal circles and divide them up, one into 3 even parts, the other into six:

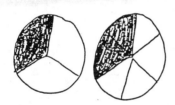

I would then shade in the desired fraction for each circle, but on the same side so that it would be very obvious that they were the same. I would then have them erase two of the lines in the non-shaded region that did not match of the circle divided into six parts that did not match with the other circle. By erasing they would see the same amount is shaded and

Begin your response to question 2 here.

the same amount is left over.

However for larger fractions, this might not prove as effective. Instead, I would use the strategy of factoring. For example, $\frac{20}{40} = \frac{1}{2}$. By factoring out ~~twenty~~ and forty, students would see that by cancellation, they end up with the same number. I would explain ~~that~~ if you multiplied $\frac{1}{2}$ by $\frac{20}{20}$, you would end up with $\frac{20}{40}$, ~~sometimes~~ so also, <u>all</u> ~~some~~ fractions are multiples of each other..

Sample 9: Score of 3

The response that begins on the next page demonstrates an understanding of the mathematics to be presented. The response only indicates (rather than demonstrates) how to present the mathematics in a way that can reasonably be expected to achieve the desired goals. Only one example is given in the entire response, and a method for determining whether two fractions are equivalent is only barely indicated (the response explains that two equivalent fractions can be changed to equal each other).

This response is almost complete and the parts of the response that are complete are well motivated.

-2-

Begin your response here.

I would definitely use pictures or pie circle-manipulative pie-pieces. Seventh graders often need concrete visual examples in order to fully grasp a concept. If manipulatives are available, they are probably preferred because they allow the students to actually physically move them.

I would probably start with 'easy' equivalent fractions like $\frac{1}{2} = \frac{2}{4}$. The manipulatives pie pieces (might not be their right name) are circles that are cut in different proportions. One might be halves, another thirds, another fourths, and so on. When looking $\frac{1}{2} = \frac{2}{4}$, take the circle that is cut is two parts and the circle cut in four parts.

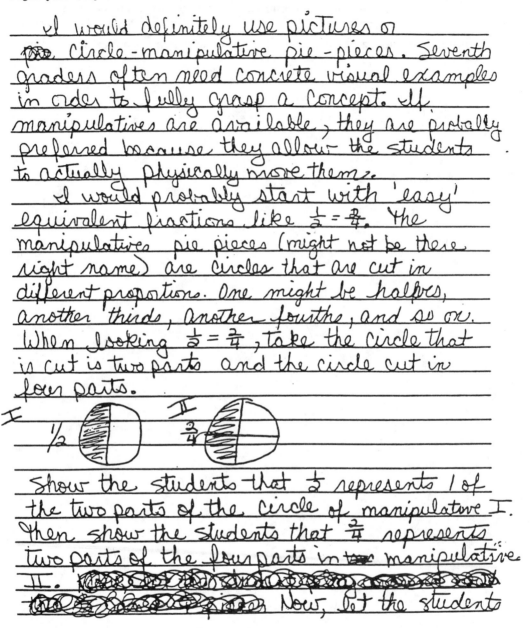

Show the students that $\frac{1}{2}$ represents 1 of the two parts of the circle of manipulative I. Then show the students that $\frac{2}{4}$ represents two parts of the four parts in the manipulative II. ~~scribbled out~~ Now, let the students

GO ON TO THE NEXT PAGE.

work with the one half piece and the two one-fourth pieces. Ask students if they see any similarities among these three pieces.

Students might suggest that $\frac{1}{4} + \frac{1}{4} = \frac{1}{2}$. Ask them to prove it with the pieces. Students should be able to recognize the fact that the 2 one-fourth pieces ~~and~~ can be placed directly one top of the one-half piece. ~~And that's to say the~~ these configurations have the same area. Ask students to conjecture about what other parts of circles or fractions might have the same results. Along students time to work with the manipulatives in pairs to test their conjectures.

After this bring all the students back together and discuss the idea of equivalent fractions ~~tw~~ represent the same area ~~or~~ amount. Then discuss the idea that equivalent fractions can be changed to actually equal each other.

ex $\frac{1}{2} = \frac{2}{4}$

$\frac{1}{2} = \frac{1}{2}$

GO ON TO THE NEXT PAGE.

-4-

~~[illegible crossed-out text]~~

The key to this lesson is allowing students to work ~~[crossed out]~~ with the manipulatives to be able to visualize that equivalent fractions are the same quantity.

GO ON TO THE NEXT PAGE.

Sample 10: Score of 2

The response that begins on the next page is incomplete in that it addresses completely the conceptual understanding of equivalent fractions but gives no indication of how to develop the ability to determine whether two fractions are equivalent. Because the response does not do so (address how to determine whether two fractions are equivalent), it also fails to demonstrate an understanding of the mathematics involved in this procedure.

Thus the response is incomplete and demonstrates only a limited understanding of the mathematics to be presented.

Question 2

Begin your response here.

One way to develop students interest in a subject is to use familiar objects to teach the concepts. Most students are familiar with how a round pizza is sliced into pieces for serving. Using large circles of poster board with either a picture of a pizza (same size as the circle) or ingredients drawn to represent each circle as a pizza could be cut into varying amounts of pieces. One pizza would be left whole, one cut into two halves, one into four fourths, and one into eight eighths. (Suggestion: laminating the pizzas would preserve the manipulatives in a better condition for future use). Smaller versions of these manipulatives could be used by the students at their desks, but only after the lesson is taught by the teacher so the students could remain focused on the lesson.

To teach the concepts, the teacher would show the students the "whole" pizza and then place it flat upon a table at the front of the room where everyone can see (or, have the students put their desks in a circle to cancel out any interference). Next, the teacher takes the two halves of the second pizza, holds them up to show the students, and then places them on top of the "whole" pizza, covering it completely. Then, the teacher follows the same procedure with the "fourths" pizza and

GO ON TO THE NEXT PAGE.

–2–

the "eighths" pizza. The teacher then can use this stack of pizzas to demonstrate the equivalent fractions of ½, ¾, and ⁴/₈ by picking up one of the "halves", leaving the fourths and eighths still on top of the half he a she picked up.

The teacher can then say that we know that the fraction ½ shows not that we have one-half of the pizza, emphasizing this by writing "½" on the board. "To find some equivalent fractions for one-half (the teacher writes an "equal" sign after the ½), let's look at our other layers of pizza, the teacher might say. The top layer is in eighths, so the teacher could have the students count off the pieces from that layer as they are removed. Since there are four eighths and this amount of pizza covered the exact same area as the one-half did, we can say that ½=⁴/₈. The teacher would then write "⁴/₈" after the equal sign.

Similarly, the teacher can perform the same operation with the "fourths" layer of pizza and add two fourths to the series of equalities to get ½ = ⁴/₈ = ²/₄.

GO ON TO THE NEXT PAGE.

-3-

Similar manipulatives can be constructed for the series $\frac{1}{3} = \frac{2}{6} = \frac{4}{12}$. If the teacher does not have the time or materials to provide manipulatives for each student, a series of work stations through which the students rotate in groups.

In making these manipulatives, it might make it easier for the students to differentiate between the different layers of stacked pizzas if the poster board backing is a different color for each level. This would also aid them in assembling the stack of pizzas and keep the "halves" series and "thirds" series from being mixed together.

GO ON TO THE NEXT PAGE.

Sample 11: Score of 1

The response that begins on the next page demonstrates a very limited understanding of the mathematics to be presented. The response indicates that $\frac{2}{3}$ may equal $\frac{4}{6}$ but makes no conclusions as to whether they are equivalent or what this equivalence would mean. Although the response demonstrates a very limited understanding of at least one example of possibly equivalent fractions, it does not demonstrate even a very limited understanding of how to determine whether two fractions are equivalent.

Begin your response to question 2 here.

PROBEEM : $\frac{2}{3} \overset{?}{=} \frac{4}{6}$

Tell the students to take 2 pies and ~~and~~ divide the ~~1pic~~ into

3 equal Parts. ~~on one of these pies,~~ and the other pie into

6 equal parts

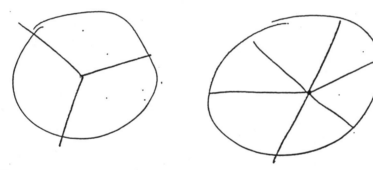

(1.) Tell the students to divide 1st pie into three equal parts and ~~to~~ tell them TO REMEMBER HOW THEY CUT THE PIEs.

(2) Using those same 3 cuts on the 1st pie, put those cuts into the 2nd pie. Then tell the students to cut the 3 slices in half.

(Question 2 continued)

EXTENSION ON 1

GIVE 2 slices OF THE 1st pie
To ~~# + member of the group~~
Group A

EXT. ON 2

Give Four slices of the second pie
to Group B

Question 3

This section begins with a discussion of the solution to Question 3, followed by several actual responses, with comments from the lead scorers explaining how the scoring guide above was used to rate each response.

Question 3–Solution

In order for the students to discover the difference between "regular" and "equilateral," they must at some point be given the definition of each of those terms. The desired investigation involves an exploration of polygons with more than three sides. Therefore the investigation should include polygons with different numbers of sides so that the students can see that the concept of "regular" and the concept of "equilateral" differ for *n*-gons for several values of *n* that are greater than 3. Finally, in order to serve its purpose as an investigation, the exercise should involve hands-on activity and exploration by students, in addition to any examples presented by the teacher.

A sample solution is shown below. The sample responses also provide good models for solutions.

I would first have the students write the definitions on the board and clarify what the difference between the two definitions is.

Equilateral – All sides of the polygon have the same length.

Regular – All sides and all angles of the polygon have the same measure.

Then I would have the students explore whether a polygon could be equilateral without being equiangular *and* equilateral (i.e., regular). I would give each student 8 strips of cardboard, all of the same length, and 8 thumbtacks to attach the strips at the vertices. Then they would be directed to make equilateral polygons with 3, 4, 5, 6, 7, and 8 sides, respectively. They should discover that they are able to construct polygons that, although they are equilateral, are not equiangular, and thus are not regular. The use of the thumbtacks at the vertices would permit the students to push nonrigid figures into different shapes with unequal angles. I would require that the students record the results of their experiments and keep them for future reference.

Question 3–Sample Responses

Sample 12: Score of 5

The response that begins on the next page clearly demonstrates an understanding of the mathematics to be presented. The math is presented to the students in a way that is likely to achieve the desired goal of discovering that not all equilateral figures are regular.

The response is clear and complete.

The response is well motivated in that the students are controlling a significant portion of the investigation and are creating figures for themselves. It is also appropriate that figures with more than four sides are investigated, so that the students can see that figures with four sides do not represent a single special case for which "equilateral" and "regular" have different meanings.

Question 3

Students in geometry frequently confuse the ideas of "equilateral" and "regular." For a triangle, these mean the same thing, but a polygon with more than 3 sides can be equilateral without being regular. Describe an investigation you would have your students make to discover the difference between regular and equilateral. Your investigation may involve the use of any type of manipulative or any software package.

NOTES

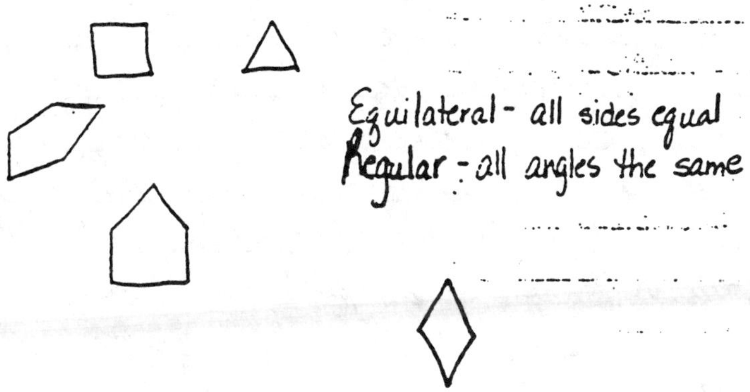

Equilateral - all sides equal
Regular - all angles the same

GO ON TO THE NEXT PAGE.

-2-

Begin your response here.

I would have the students define equilateral
and regular. Equilateral means that all sides
are equal and regular means that all sides
and all angles are equal. Then I would
divide the students into groups (3 or 4). In
each group, I would assign a type of polygon,
one group having quadrilaterals, one group
having pentagons, one group having hexagons
etc. I would give each group the appropriate
number of popsicle sticks according to the
number of sides that their polygon has.
I would then have them try to make
polygons that are not regular. Since the
popsicle sticks will all be of the same
length, the polygons must be equilateral.
So, if they can create a polygon that is
not regular then they have shown that
equilateral and regular are not one and
the same. Then the groups would discuss
the differences between regular and
equilateral. Finally, each group would
share their results with the class,
reinforcing the discovery made.

GO ON TO THE NEXT PAGE.

Sample 13: Score of 4

In the response that begins on the next page, although the definitions of "equilateral" and "regular" are not given explicitly, the response clearly demonstrates an understanding of the mathematics to be presented and how to present it in a way that is likely to produce the desired result of discovering that equilateral polygons are not necessarily regular.

The response is complete and fairly well motivated in that the students are controlling a significant portion of the investigation and are creating figures for themselves. The investigation could be better motivated if it specifically indicated that students would be asked to examine figures with more than four sides in order to see that the properties being discovered do not hold for quadrilaterals only.

Begin your response to question 3 here.

I would bring to class some tape and a box
of straws. I would have the students construct a
triangle by taping the ends of 3 straws together

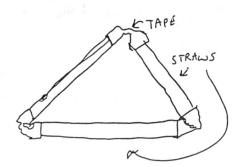

I would Have the students compare their triangles.
They should notice that the angles are all equal and
the sides (straws) are all the same length. So it
is equilateral AND REGULAR.

Then I would have them make A figure with more than 3
sides, using again straws AND tape. The students should
see that the sides may all be equal, but the figure
is more manipulative than the triangle. They can make
an equilateral figure without having all equal
angles.

Sample 14: Score of 3

The response that begins on the next page demonstrates an understanding of the mathematics to be presented. This understanding is demonstrated less clearly than in responses receiving a higher score because the response never clearly indicates the meaning of "equilateral," either implicitly or explicitly.

The response is only fairly well motivated in that the students are only measuring figures given to them and not creating any figures of their own. This gives the activity more of the flavor of an exercise than an investigation or discovery.

The response indicates that the activity to be undertaken by students will consist of measuring sides and angles and recording the measurements. The response does not demonstrate how this will lead the students to the appropriate conclusions. Thus, this is also a response that indicates how to present the material to the students in a way that is likely to achieve the desired result but that fails to demonstrate how to present the material.

-2-

Begin your response here.

  I would explain that in order for a polygon to be regular it must have equal sides and equal angles. I would then give the students several figures and a ruler and a protractor. All of the figures will be equilateral, but all won't be regular.

  I will have the students measure each side & angle of each figure and record their measurements.

  This activity should cause them to remember that just because the sides are equal, does neccessary mean that the polygon will be regular.

GO ON TO THE NEXT PAGE.

Sample 15: Score of 2

The response that begins on the next page demonstrates only a limited understanding of the mathematics to be presented. The response does not indicate an understanding of the meaning of "regular polygon" or "equilateral polygon."

The response is also somewhat unclear in that the exercise described seems designed to produce a definition of "regular," rather than to discover that the two definitions for "equilateral" and "regular" are not equivalent.

Begin your response to question 3 here.

I would divide the class, randomly, into small groups of 3 to 4 students. Each student would be assigned a job within the group, and each group would be given an assignment sheet. The question they would be instructed to answer would be "what is a regular polygon?" and "How is it different from an equiangular polygon?" The assignment sheet would have the following assignment written on the top:

"In the front of the room are two boxes, one labeled regular, one labeled equilateral. Send the <u>gopher</u> from your group to get one polygon from each box. The <u>recorder</u> should then draw the polygons on the assignment sheet under columns labeled regular and equilateral. The <u>measurer</u> should then begin measuring each polygon. Measure all angles and line segments. Repeat this exercise 3 to 5 times. As a group formulate a definition for regular. When you are finished think about and try to answer the question, "what is the largest regular office building in the country?"

This group work will be collected and followed up the next day by a discussion

Sample 16: Score of 1

The response that begins on the next page demonstrates a very limited understanding of the mathematics to be presented. Little or no effort is made to describe an investigation that is likely to achieve the desired goals.

Question 3

Begin your response here.

Give many examples of regular and equilateral polygons on the overhead projector. Beside it give many examples of equilateral and non-regular polygons. Repeatedly state the definition of an equilateral polygon "all sides are equal." Let the students discover the differences between the two sets. This should allow them to come up with the difference and therefore the understanding between "equilateral" and "regular."

GO ON TO THE NEXT PAGE.

Chapter 12
Are You Ready? Last-Minute Tips

► ► ► ► ► ► ► ► ► ► ►

Checklist

Complete this checklist to determine if you're ready to take your test.

❏ Do you know the testing requirements for your teaching field in the state(s) where you plan to teach?

❏ Have you followed all of the test registration procedures?

❏ Do you know the topics that will be covered in each test you plan to take?

❏ Have you reviewed any textbooks, class notes, and course readings that relate to the topics covered?

❏ Do you know how long the test will take and the number of questions it contains? Have you considered how you will pace your work?

❏ Are you familiar with the test directions and the types of questions for the test?

❏ Are you familiar with the recommended test-taking strategies and tips?

❏ Have you practiced by working through the practice test questions at a pace similar to that of an actual test?

❏ If you are repeating a Praxis Series™ Assessment, have you analyzed your previous score report to determine areas where additional study and test preparation could be useful?

The Day of the Test

You should have ended your review a day or two before the actual test date. And many clichés you may have heard about the day of the test are true. You should

- Be well rested.

- Take photo identification with you.

- Take a supply of well-sharpened #2 pencils (at least three) for a multiple-choice test. You have a choice of using pencil or blue or black ink pen for a constructed-response test. Take an adequate supply of your choice.

- Eat before you take the test, and take some food or a snack to keep your energy level up.

- Be prepared to stand in line to check in or to wait while other test takers are being checked in.

You can't control the testing situation, but you can control yourself. Stay calm. The supervisors are well trained and make every effort to provide uniform testing conditions, but don't let it bother you if the test doesn't start exactly on time. You will have the necessary amount of time once it does start.

You can think of preparing for this test as training for an athletic event. Once you've trained, and prepared, and rested, give it everything you've got. Good luck.

Appendix A
Study Plan Sheet

Study Plan Sheet

See Chapter 1 for suggestions on using this Study Plan Sheet.

STUDY PLAN						
Content covered on test	How well do I know the content?	What material do I have for studying this content?	What material do I need for studying this content?	Where could I find the materials I need?	Dates planned for study of content	Dates completed

Appendix B
For More Information

▶   ▶   ▶   ▶   ▶   ▶   ▶   ▶   ▶   ▶   ▶   ▶

ETS offers additional information to assist you in preparing for The Praxis Series™ Assessments. You can also obtain more information, and review *Tests at a Glance* by visiting our Web site: www.ets.org/praxis.

General Inquiries

Phone: 800-772-9476 or 609-771-7395 (Monday-Friday, 8:00 A.M. to 7:45 P.M., Eastern time)
Fax: 609-771-7906

Extended Time

If you have a learning disability or if English is not your primary language, you can apply to be given more time to take your test. The *Registration Bulletin* tells you how you can qualify for extended time.

Disability Services

Phone: 866-387-8602 or 609-771-7780
Fax: 609-771-7906
TTY (for deaf or hard-of-hearing callers): 609-771-7714

Mailing Address

ETS–The Praxis Series™
P.O. Box 6051
Princeton, NJ 08541-6051

Overnight Delivery Address

ETS–The Praxis Series™
Distribution Center
225 Phillips Blvd.
Ewing, NJ 08628-7435